Palgrave Studies in the History of Science and Technology

Series Editors

James Rodger Fleming
Colby College
Waterville, Maine, USA

Roger D. Launius
Smithsonian Institution
National Air and Space Museum
Washington, D.C., USA

Designed to bridge the gap between the history of science and the history of technology, this series publishes the best new work by promising and accomplished authors in both areas. In particular, it offers historical perspectives on issues of current and ongoing concern, provides international and global perspectives on scientific issues, and encourages productive communication between historians and practicing scientists.

More information about this series at
http://www.springer.com/series/14581

Mark Monmonier

Patents and Cartographic Inventions

A New Perspective for Map History

palgrave
macmillan

Mark Monmonier
Department of Geography,
Maxwell School of Citizenship and Public Affairs
Syracuse University
Syracuse, New York, USA

Palgrave Studies in the History of Science and Technology
ISBN 978-3-319-84551-7   ISBN 978-3-319-51040-8   (eBook)
DOI 10.1007/978-3-319-51040-8

Cover image Herman E. Schulse
Cover design by Fatima Jamadar

Printed on acid-free paper

This Palgrave Macmillan imprint is published by Springer Nature
The registered company is Springer International Publishing AG
The registered company address is: Gewerbestrasse 11, 6330 Cham, Switzerland

To the memory of my maternal grandfather,
Wesley W. Mason (1879–1932),
whom I never met. He held a variety of jobs (including private detective
and plant foreman) before becoming a full-time inventor and was awarded
several patents for bottle caps and crown-making devices.

PREFACE

This book explores an arena of cartographic creativity largely ignored by map historians: the patent system, whereby an inventor can lay claim to a novel idea and control its use for two decades. As I argue in Chap. 1, the patent system is not just a way to get ideas in print but also a parallel literature, similar in fundamental ways to the conventional academic-scientific-technical literature of books and journal articles. Although the patent system appeals to a different kind of innovator—someone with a product in mind and a decidedly more practical bent than the typical scholar—it is a coherent literature, with a vetting process, distribution channels, citation protocols, and searchable databases. In this milieu the patent examiner serves as both editor and peer reviewer, and the vetting, as I show, can be contentious and protracted. Although patents are characterized by a distinctive jargon I call patentese and by a heavy reliance on drawings to explain a device or process, the published patent, like the published journal article, addresses a shared need for achievement that motivates inventors and scholars alike.

Unlike most map histories, this book is more about devices and techniques than about maps and atlases. Cartographic inventors who filed patents created clever products and processes intended to make map information easier to use—a cartographic variant of the better mouse-trap. Their emphasis on the quotidian is apparent in the book's individual chapters, which focus on admittedly mundane applications like streetcar transfers, rural address guides, mechanical route-following machines that anticipated the GPS, folding schemes, world map projections, and diverse improvements of the terrestrial globe. The final chapter, with an emphasis on you-are-here mall maps, relates early applications of electricity in interactive mapping to the rise of digital cartography and the emergence of patent trolls, for whom the patent is largely a license to litigate.

Coverage begins in the mid-nineteenth century and extends through the pre-computer era, with additional narrowing in individual chapters as developments dictate. Except for noting the emergence of digital cartography in the final chapter, I avoid the complications of semiconductors and software, which demand a book of their own. In addition, my focus on patents awarded in the United States reflects a dearth of data on applications abandoned or rejected as well as the complexity of exploring patents issued in Britain, Canada, France, Germany, and Switzerland, among others. By concentrating on US patents, I could rely on complementary online databases maintained by the US Patent and Trademark Office and by the Google Patents Project as well as pre-1975 case files in the National Archives at Kansas City. But because of an increasingly international market for intellectual property, non-US residents figure prominently in the chapters on map folding and map projections.

In seeking insights and understanding, I employed a four-pronged research strategy that began by using the US Patent Classification System to identify relevant patents, principally in the Printed matter/Maps and Education and demonstration/Geography categories. Google Patents provided PDF files of published patents, including all artwork. Case files from the National Archives, which include correspondence between the Patent Office and the inventor and the inventor's attorney, shed light on the examiner's concerns, including an effort to both clarify and narrow the inventor's claims. For added background on the inventors, I relied on genealogical research tools, principally Ancestry Library Edition, which provided a generally useful portal to manuscript census schedules, city directories, draft board records, and similar components of Big Microdata, which sometimes yielded key details about an inventor's education, training, and relevant family connections. I say "sometimes" because the inventors varied widely in their biographical footprints, which online searching of newspaper databases and the HathiTrust Digital Library could occasionally clarify—advertisements and news articles, where available, were valuable evidence of commercial follow-through. Google Patents also provided useful details about an inventor's prior or subsequent experience with the patent system. Finally, I tabulated counts of inventions by year and category, used simple statistical analysis to compare trends in map-related patents with patenting in general, and created time-series frequency plots, all of which enriched the narrative with an analytical framework. If pressed to sum up my research design in a few words, I would call it, simply, archival and statistical graphics.

How one writes is no less important than how he (or she) looks for and interprets data. Creating a narrative that is at once informative, incisive, reliable, and readable requires some cherry picking. That said, in all chapters, the most important inventions and inventors were impossible to avoid, hence the inclusion of a few inventors with frustratingly faint biographical footprints. In some cases, I might have said more, but chose not to, in the interest of concision and narrative flow: these chapters are stories and I wanted to make them as coherent as the facts allow. In other cases, I might have said less but was unable to resist an intriguing tidbit that adds, I hope, to the reader's understanding of the complexity of the patent system as a parallel cartographic literature.

Constructing the narrative was often an exercise in cautious interpolation when connecting facts (some firmer than others) and discerning motivations in the absence of inventors' diaries, business records, and personal correspondence. My reliance on inference when stitching together an explanation should be apparent in subtle caveats like *plausible*, *perhaps*, and *would have*.

Frequent direct quotations from a published patent as well as from correspondence in its case file are intended to convey a flavor of not only the jargon I call patentese but also the patent examiner's careful (and occasionally cantankerous) scrutiny of the inventor's claims. The inventor's, examiner's, or attorney's original wording is often sufficiently concise to obviate a paraphrase.

Because the flavor of an invention is frequently best communicated by the inventor's drawings, my narrative also incorporates numerous facsimiles to help explain how a device or process works, to show the inventor's concern for detail, and to illustrate the patent's distinctive form as a geometrically rich literary text. As stark abstractions, these diagrams attest to the difficulty of describing an innovation with words alone—explaining it to a manufacturer who might want to license it, describing it to a competing inventor who needs to know what's "old" and thus no longer patentable, and informing an attorney eager to resist a patent examiner's objections or fend off a charge of infringement.

Mark Monmonier
Syracuse, NY

ACKNOWLEDGMENTS

During this book's concentrated two-year period of gestation, I depended on the support of many institutions and people. The Maxwell School of Citizenship and Public Affairs at Syracuse University provided a one-semester research leave; the Geography and the Spatial Sciences (GSS) Program and the Science, Technology, and Society Program (STS) at the National Science Foundation supported a graduate research assistant for one year; and the National Archives at Kansas City provided access to the case files for numerous patents. I am especially indebted to Jake Ersland and Robert Beebe, archivists in Kansas City; David Delzingaro, in the editorial division of the US Patent and Trademark Office; and Ralph Ehrenberg, of the Geography and Map Division of the US Library of Congress (and several decades ago, the National Archives). Librarians who were especially helpful include John Olson and Darle Balforte, at Syracuse University; Jenny Marie Johnson, at the University of Illinois at Urbana-Champaign; and Meredith Anne Weber, in Special Collections at The Pennsylvania State University. Eric Anderson, of the Cartography and Geographic Information Society, and Paul Young, at the US Geological Survey, provided valuable information on John Snyder, and Maureen Reynolds, Tompkins County Clerk, assisted with corporate records. Claudia Asch, who worked with me on the History of Cartography Project, helped with German translation; Christina Leigh Dietz, of the Maxwell School, provided valuable guidance on the NSF grant; and Emily Bukowski and Christopher Robert Allen assisted in ferreting out relevant facts about several of the inventors. Joe Stoll, staff cartographer in the SU geography department, provided valuable advice on Photoshop; Brian von Knoblauch, Mike Cavallaro, Mike Fiorentino, and Stan Ziemba, all in the Maxwell School's Instruction and Computing Technology (ICT) group, helped me navigate the shoals of software and operating system upgrades and other electronic hazards; and Margie Johnson, administrative assistant in the Geography Department, helped me cope with the arcana of the university's accounting systems. Many thanks too to my wife Marge for encouragement and constant support.

CONTENTS

LIST OF TABLES

CHAPTER 1

Maps and Patents

On 17 October 1916, the US Patent Office awarded patent 1,201,605, titled "Transfer-Ticket," to Albany, New York, residents William C. Moffatt and Arnold von Schrenk.[1] The published patent included a single image: a hand-drawn prototype of a streetcar transfer with a small schematic map of inner city routes and transfer points (Fig. 1.1). The inventors based their example on Albany's Pine Hills line, connecting downtown with a western suburb. Horizontal and vertical lines represent key streets, and north lies to the left to fit the elongated slip of paper attached at the short end to a small pad for quick tear-off whenever a boarding passenger greeted the conductor with "Transfer please!"

Here's the invention's context. Early twentieth-century streetcars typically carried a two-person crew: a motorman to run the car and a conductor to collect fares and hand out transfers, which allowed passengers to continue their journey on a second (or perhaps even a third) line.[2] The conductor used a hand-held hole punch to mark the transfer with a distinctive cross, plus-sign, star, or circle. A punch in the table to the right, which divided 12 hours into 10-minute intervals, recorded the boarding time, and additional punches indicated boarding and transfer points. Trolley companies were leery of "looping" by passengers who tried to end a trip close to its starting point without paying an additional fare.

An annotated example (Fig. 1.2) for a hypothetical trip that began at 8:20 a.m. shows how the transfer worked. We know it's before noon because the conductor has torn off the PM stub, joined along a perforated

© Mark Monmonier 2017
Mark Monmonier, *Patents and Cartographic Inventions*,
Palgrave Studies in the History of Science and Technology,
DOI 10.1007/978-3-319-51040-8_1

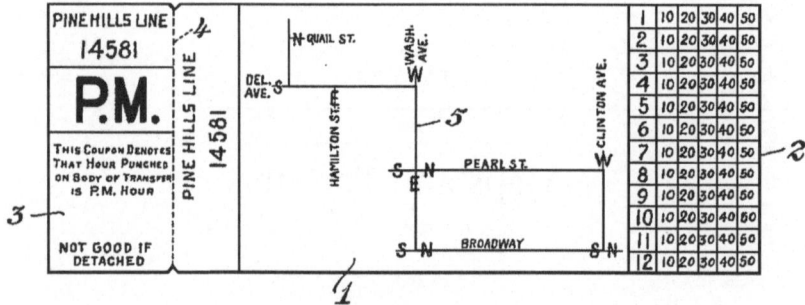

Fig. 1.1 Drawing of a streetcar transfer with a map invented by William C. Moffatt and Arnold von Schrenk (US Patent 1,201,605; 1916)

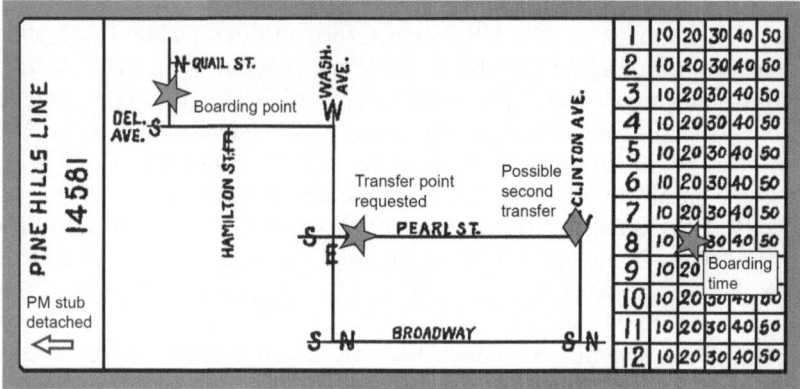

Fig. 1.2 Slightly enlarged version of the drawing in Fig. 1.1 showing holes punched for a hypothetical journey downtown by a passenger who boarded between Quail Street and Delaware Avenue at 8:20 a.m. and asked to transfer to a northbound car at Pearl Street. Detachment of PM stub made the ticket invalid that afternoon or evening. In this example, the receiving conductor on the Pearl Street line punched the transfer again so that the passenger could take a westbound car at Clinton Avenue

line. Boarding time was important because the typical trip, including connections, took less than an hour, and extended stopovers were discouraged. The conductor also punched the transfer at the upper left to show that the trip began between Quail Street and Delaware Avenue and once more, over the S at Pearl Street, because our imaginary passenger asked to transfer to a southbound car on Pearl. The receiving conductor might punch the ticket again, perhaps over the W toward the right if the passenger wanted to transfer a second time, to a westbound car along Clinton Avenue. The small, abstract map records the trip's geography and limits the transfer privilege.

Punchable transfers were not new, nor were tiny schematic maps, but their juxtaposition apparently met the Patent Office's standard of novelty and usefulness. As the patent application asserted, passengers could easily visualize connections and acceptable transfer directions, and a marked boarding point and preselected transfer direction helped the receiving conductor spot an improper round trip.

The patent gave Moffatt and von Schrenk the right to both control and profit from their invention for the next 17 years.[3] Their patent was a utility patent, the most common type, which grants fuller rights and has a longer term than a design patent, which is concerned with how an invention looks rather than how it works. Utility patents are anchored by a series of claims that collectively describe the breadth of the patent holder's monopoly. All six claims for the "Transfer-Ticket" begin with the words, "A transfer ticket having a diagrammatic representation of connecting railway routes produced on a face thereof." The claims note that the diagram is "adapted to be punched or marked" and link the ticket's validity for a "continuing trip" to "predetermined points," "trip and direction of travel," a "point of connection," or a "transfer point." The goal is to cover all plausible variations without making the patent so broad that it's easily overturned in court.

Inventors seldom write this way, at least not without help. Neither Moffatt nor von Schrenk had formal legal training, and only von Schrenk understood the streetcar industry sufficiently well to think their invention was marketable. Aside from similarity in age and place of residence and an appreciation of conveyances that run on rails, they seem unlikely business partners.

Genealogical research tools—more on these in Chap. 2—suggest that Moffatt, in his late 20s when the patent application was filed on 3 March 1913, might have lived in Albany only a year or two before moving back to Scranton, Pennsylvania, where he had worked previously as a clerk for the Delaware, Lackawanna & Western Railroad.[4] A 1914 Albany city directory reported him as "rem [removed?] to Scranton Penn," and a 1916 Scranton directory listed him as a general storekeeper for the Delaware and Hudson Coal Company, a subsidiary of the Delaware and Hudson

Railway, headquartered in Albany. Moffatt had grown up in Scranton, and electronic archives captured from local newspapers of the era suggest he was a model citizen who married the former Ruth Lumley in 1924, fathered sons in 1925 and 1927, and served as foreman of a federal grand jury in 1955.[5] He was also a pallbearer at several funerals, including one on 6 August 1934, the day before busses replaced streetcars on Scranton's Throop—Olyphant line.[6] I have no idea whether he was among the hundreds who rode one of the last cars, but it's inviting to envision a trolley buff, newly resident in Albany in the 1910s, seeing the need for an improved transfer.

Arnold von Schrenk's connection to the "Transfer-Ticket" patent is more straightforward: in 1913 he was the 32-year-old general superintendent of the United Traction Company, which ran Albany's streetcars.[7] He had graduated from Columbia University in 1902 with a degree in electrical engineering, and left Albany in late 1914 for St. Louis, to work with his older brother, Herman, a plant pathologist who consulted with railroads and utilities on the use of creosote. In the final months of World War I, Arnold was a first lieutenant in the Army Signal Corps, a further flourish on a technologically impressive resume.

Patent documents typically name the inventor's attorney, who not only advised on patentability but also served as a ghostwriter, to draft a general description, called the *specification*, and one or more protected ideas, called *claims*. The attorney also fielded pointed questions from examiners at the Patent Office.[8] Although the published patent reproduces the signatures of both the inventor and the attorney, only the inventor's name is spelled out. I struggled to interpret the name of the attorney who had signed the "Transfer-Ticket" application until I realized that "Attorneys" printed below the signature was plural, which suggested an ampersand, which helped me decode "Briesen & Knauth." For confirmation, I checked the 1912 edition of *Hubbell's Legal Dictionary for Lawyers and Businessmen*, which placed the firm in Manhattan, at 25 Broad Street, and listed five attorneys: three Briesens, one Knauth, and Otto von Schrenk, who was also a notary public. Genealogical research revealed that Otto, like Herman, was an older brother of Arnold, the inventor.

Was Otto the source of Arnold's unwarranted hope that his and William Moffatt's invention would succeed? I say unwarranted because moderately exhaustive searching failed to find a single streetcar transfer, for United Traction or any other line, based on their patented "Transfer-Ticket."[9] I turned up several images of United Traction transfers, but none with a

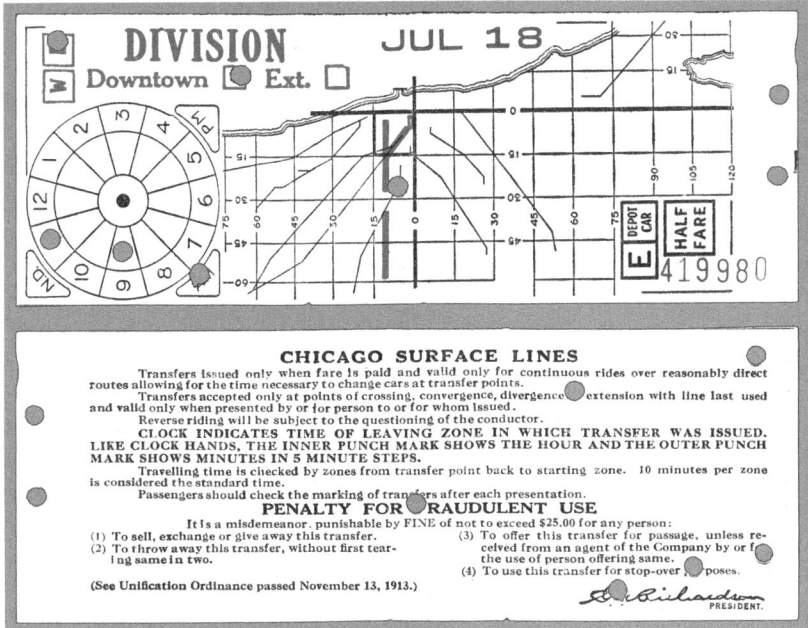

Fig. 1.3 Front (*top*) and back (*bottom*) views of a streetcar transfer used circa 1915 on Chicago's Division Street surface line. The thick dark gray vertical line that bifurcates near the lake was printed in red to mark the Division Street route. Author's collection

map. Perhaps Arnold didn't remain in Albany long enough to try it out at the company he ran.

Online searching turned up an arguably superior cartographic transfer, nicely adapted to Chicago's gridded street pattern and more complex streetcar network (Fig. 1.3). Its basic framework, printed in black, includes a clock face for recording the trip's start time and a simple map with north to the left, the shoreline of Lake Michigan at the top, and thin lines 15 blocks apart that divide the city into grid cells. Thick lines representing the grid's axes, State and Madison streets, meet downtown at the Loop, and additional lines show several carlines headed downtown. A red overprint gives the date (18 July), names the carline (Division), and plots a route that splits into Downtown and Ext[ension] destinations as it approaches the lake. Five circular holes punched on the left describe an eastbound trip that began around 9:55 a.m. on an eastbound car marked "Downtown," and an additional punch shows the boarding point. I bought this transfer on eBay, to see whether it mentioned a patent number. It didn't.

Not all patentable inventions are patented. Why pay a lawyer to draw up the application when there's little need to ward off a competitor who might exploit the idea or little likelihood of selling or licensing the patent? I suspect that many inventors file for patent protection because they revel in owning an idea, however minor, or in merely claiming priority of invention—being first can be a strong motive.[10] While there is no way to prove the "Transfer-Ticket" was a vanity patent, it's tempting to ask whether some inventors apply for patents simply because they craved recognition, or because a brother with a law degree helped them negotiate the bureaucratic maze.

Arnold's success in patenting the "Transfer-Ticket" might partly explain his filing additional patent applications between May 1916 and February 1917. By August 1919, he was the sole rights holder for three very different inventions: a removable running board for automobiles, an improved linkage for automobile and railway brakes, and a railroad spike that held tight without splitting the wood fibers in a railway tie. I have no idea whether any of these inventions ever went into production. The law firm named on the third patent was Briesen and Schrenk, recently renamed with Otto von Schrenk in second place on the letterhead. That Arnold's other two patents list other law firms indicate that he was not totally dependent on Otto.

However inconsequential its impact, the "Transfer-Ticket" is part of the trove of published cartographic patents that constitutes what I call a parallel literature, akin to the standard literature of academic-scientific-technical cartography as a discourse for sharing information and documenting progress. Despite obvious structural differences between scholarly publishers and the US Patent Office (renamed the Patent and Trademark Office in 1975), the patent system is fundamentally similar to peer-reviewed journals insofar as both endeavors promote quality control and disseminate worthy ideas. Indeed, both types of publishing require a conscientious vetting of manuscripts, for which scholarly journals rely upon blind review by well-qualified experts, whereas the Patent Office employs experienced examiners trained in law and engineering, who take on the roles of editors and referees. Patent examiners can be more proactive than their academic counterparts insofar as they typically provide most, if not all, the citations.[11] Although the standards of usefulness are fundamentally different—a scholarly article respected for its philosophical or rhetorical impact need not improve the way we make or use maps—both arenas demand creativity and originality. By analogy, an insightful critique of another scholar's work is akin to a marked improvement of an existing invention. Progress in scholarship and patented invention is largely incremental.

By any definition, the patent system is a distinct (and distinguished) literature. Every approved application is given a unique patent number along with a brief title and issue date and is circulated in print to promote what the US Constitution calls the "useful arts."[12] Mandated disclosure lets manufacturers know about new ideas available for sale or license and tells competing inventors to focus elsewhere unless they can conjure up a further improvement not covered by the patent holder's claims.[13] Although the early decades of patent registration in the United States were haphazard and understaffed—the federal government didn't move from Philadelphia to Washington until 1800[14]—the Patent Office began publishing descriptions of patents in its annual report in 1843, the complete specifications and drawings of individual patents in 1871, and shorter descriptions in the *Official Gazette of the United States Patent Office*, issued weekly since 1872. The *Gazette* also includes summaries of court decisions, changes in procedure, and other news, and its linage is carefully rationed. Although the full patent for the "Transfer-Ticket" ran to four pages, the abridged version, which occupied less than a full page in the *Gazette*, included only five of the six claims and a photoreduced version of the single drawing.[15] A 1924 report by the Institute for Government Research found that "the Patent Office spends more for printing than any other bureau of the national government."[16] Moreover, 50 of the 97 linotype machines at the Government Printing Office were reserved for Patent Office work.

Images are an essential part of any patent. By convention, these images are line drawings, usually drafted in pen-and-ink and lettered by hand, as in Fig. 1.1. And because the drawings require explanations, numbers link particular features to the accompanying text. Although complex inventions might require more than a hundred numbers distributed over more than a dozen drawings, the "Transfer-Ticket" required only five numbers to point out "the main or body portion 1," "the portion 2 having a main column of figures indicating the twelve hours of the day," "the p. m. coupon 3 ... attached to the body 1 by means of a weakened line 4," and "a map 5 or diagrammatic representation of connecting railway routes." Because these drawings provide a concise overview of an invention, they precede the patent's verbal description. When a patent has more than one drawing, the image labeled "Fig. 1.1" is always the most representative drawing and the obvious candidate for the single illustration that introduced the abbreviated description in the *Official Gazette*.

Historian William Rankin, who explored the intent and evolution of patent drawings, argues that they served the rhetorical role of communicating the applicant's claims to a hypothetical reader "skilled in the art."[17] Intended to be read at a glance and readily reproduced with photolithography, these black-and-white drawings were a "universal visual language" that was the visual counterpart of verbal legalese insofar as they could disclose an invention's novelty as well as claim an intellectual property right in ways not possible with words alone. Stylistic conventions such as shading and shadows that promoted legibility began to change in the 1960s, and present-day patent drawings are no longer so readily understandable to the educated lay reader.

That the numerous patent drawings in this book can be used without requesting permission or paying a fee underscores a key difference between patents and copyrights. Although a patent restricts the use of a process that is practical, novel, and not obvious, the patent holder benefits from this monopoly only when competitors or potential buyers or licensees know about the invention. Mandated publication is thus an indispensable part of the patent system. By contrast, a copyright restricts the reproduction of a creative literary, musical, or visual work so that the copyright holder can earn royalties through the sale of copies. Although a copyright originally lasted only fourteen years, with the possibility of one renewal, the term has grown to 120 years or more.[18] Because images in patent applications are not copyright protected, they can be used freely in books and articles from day one.

The notion of parallel literatures underscores the typically minimal interaction between cartographic scholars who publish in academic journals and inventors of map-related contrivances who publish through the patent system. I became more fully aware of this gap while editing *Cartography in the Twentieth Century*, the million-word encyclopedia published in 2015 by the University of Chicago Press as Volume 6 of the *History of Cartography*. As volume editor, I had ready access to full-text files for the 529 entries written by our 323 contributors and co-contributors, mostly academics or government scientists, who used *patent* as a noun, verb, or adjective only 59 times. Eighteen of these mentions are in the entry "Intellectual Property," which includes 84 instances of *copyright*—a topic that accounts for 33 of the entry's 42 paragraphs, only three of which discuss patents.[19] What's more, individual patents account for only five of the volume's 5115 references: just a tenth of a percent. Clearly, cartographic scholars have had little interest in citing patented inventions.

Similarly, the scarcity of academic articles referenced in patents suggests that inventors interested in maps have little appreciation of cartographic scholarship, even writings focused on the development and evaluation of mapping techniques. When a patent refers to a prior development, it usually just cites another patent.[20]

Negligible cross-citation between the two discourses reflects separate networks of creative people who differ markedly in work environment and sense of problem. Although cartographic inventors who rely on the patent system are hardly a coherent group, their published descriptions and claims reveal a strong appreciation of practical considerations, such as convenient compactness—a widespread need ignored by academic cartographers, whose writings reflect little interest in how to fold a map. This bias against everyday utility underscores a philosophical divide between inventors seeking an intellectual-property right and researchers seeking improved understanding of how maps work as well as recognition from fellow scholars. A breakthrough innovation ignored by academic geographers and professional mapmakers of the early twentieth century are several patents examined in Chap. 3 that anticipated by more than 50 years the in-vehicle navigation system known as the GPS or satnav.

Another prominent difference between the two literatures is the grouping of cartographic articles into comparatively homogeneous journals identified by names like *Cartographica* or *Surveying and Mapping* and readily mined for periodicals indexes or specialized bibliographies like *Bibliographia Cartographica*, which make it easy to find cartographic articles on topics like line generalization or visual variables. By contrast, cartographic patents are lumped together with other inventions in a motley mix in which numerical sequence signifies little more than a shared publication date. It's hardly surprising that the "Transfer-Ticket," published as patent number 1,201,605, has nothing in common with numerical neighbors titled "Equalizer for Vehicle Springs" and "Dental Casting-Machine," which precede and follow it in the *Official Gazette*. The most efficient way to find cartographic patents is through the United States Patent Classification (USPC) system and similar schemes devised in Canada, Europe, and Japan to help patent examiners search for "prior art" (*art* as in "state of the art"), that is, evidence that an invention is not novel.

Technological progress, which expands the possibilities for invention and alters the standards for novelty, requires an ongoing expansion and continual reshuffling of the categories used to pigeonhole inventions. The first official classification was introduced in 1830, when 6170 patents were grouped into 16 categories. As the nineteenth century unfolded, the number of classes recognized by the US Patent Office grew from 22 in 1836 to 145 in 1872 and 226 in 1897.[21] Revision and reclassification had become burdensome responsibilities by 1898, when Congress established a separate Classification Division to manage the proliferation of subclasses needed to further refine the categories.[22] A 2012 report noted that the USPC had grown to more than 450 classes with more than 150,000 subclasses.[23] In 2015, the system was officially mothballed—"relegated to a static art collection," I was told—after US patent officials joined with their European counterparts to adopt a global standard called the Cooperative Patent Classification (CPC) system.[24]

The "Transfer-Ticket" patent reflects this hierarchy of class and subclass. In 1916 the *Official Gazette* reported its assignment to "Cl. 11—15," which a 1916 edition of the Patent Office's *Manual of Classification* decodes as "Bookbinding—Tickets," a class/subclass juxtaposition that reflects the stapling together of printed streetcar transfers into small books, much like checks and postage stamps, which were separate subclasses within the bookbinding category.[25]

Fortunately for my research, the Patent Office recognized that many inventions do not fit conveniently in a single category. For example, the "Transfer-Ticket" was assigned to three USPC class/subclass categories: **283/100**, 283/105, and 283/34—the number following the slash identifies a specific subclass. In this schema the first category, in boldface, is the "original category," or OR, and the other two are "cross reference classifications," or XRs. All three belong to the "Printed matter" class (283), established in 1918. As described in staid patentese on the USPTO's classification website, category 283/100, the OR for Moffatt and von Schrenk's invention, is a first-level "indented" subclass titled "By removable material" and situated beneath a "mainline" subclass described as "Having revealable concealed information, fraud preventer or detector, use preventer or detector, or identifier" (283/72)—a wordy reference to the system of punches intended to discourage looping.[26] The two XRs are more straightforward: 283/105 is a second-level indented subclass described as "Perforated," and 283/34 is a mainline subclass titled "Maps." Without this second cross-reference, I would never have discovered the "Transfer-Ticket."

Category 283/34 (Printed matter/Maps) and its first-level indented subclass 283/35 (Printed matter/Indexed maps) are an ideal starting point for exploring cartographic patents. The USPC assigned inventions to 283/34 because their "indicia delineate geographical features," and to 283/35 because of "indicia involving means which facilitate finding some of the geographical features." (*Indicia* is patentese for defining characteristics.) These two subclass definitions match the conventional notion of *map* as well as most definitions found in cartographic textbooks.[27] These categories seemed a good place to start even though map-related inventions can be found in other categories, such as the class "Education and demonstration" (434), established in 1980. Its mainline subclass "Geography" (434/130), for "Subject matter relating to the study of features of an area of the earth," has 23 subordinate subclasses, including "Terrestrial globe or accessory therefor" (434/131) and "Relief globe" (434/132), the subject of Chap. 6.

I began my systematic exploration of cartographic patents by downloading the numbers, names, and issue dates of all patents assigned to categories 283/34 and 283/35, which I'll refer to simply as "maps" and "indexed maps." These two categories yielded a database of 304 unique patents issued between 1840 and 2012. I say unique because 26 of them were assigned to both categories. I also downloaded full-text scans of the published patent applications from Google Patents, a massive digitizing and database project useful to inventors, scholars, and the company itself—like Microsoft, Apple, and other technology giants, Google has acquired a multitude of patents, which made Google Patents a logical extension of its Google Books and Google Scholar projects.[28] These published patents are my source for filing dates, which are more relevant for historical analysis than issue dates, which can lag behind the application date by a few months or more than a decade.[29]

To explore temporal trends in cartographic patents, I constructed a time-series plot (Fig. 1.4), which shows two roughly 20-year periods of comparatively high activity and two longer stretches of relatively low, intermittent filings. Because the two earliest cartographic patents in my dataset, issued in 1840 and 1866, lacked filing dates, the graph begins with a patent filed in 1874 by Charles Waite, a Chicago engraver, and issued the same year for an invention titled "Improvement in Tickets"—each of its six examples describes opposite sides of a railroad ticket on which the front identifies the trip's origin and destination, and the back is a small map of intermediate stations and connections (Fig. 1.5).[30] At the time line's far

Patents filed, by year, 1874 to 2009, for maps (283/34) and indexed maps (283/35)

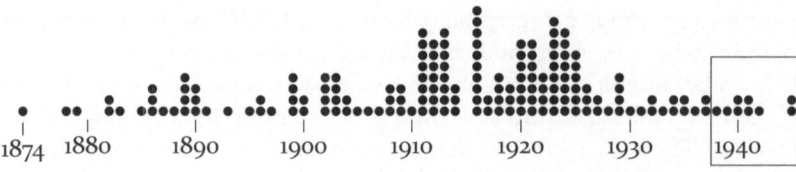

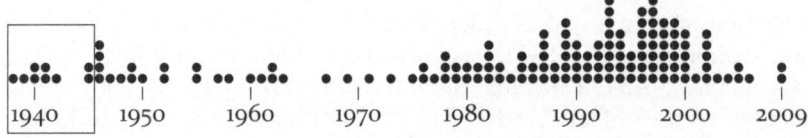

Fig. 1.4 Time-series plot of map and index map patents, by year of filing, 1874–2009. Each dot represents one patent. Splitting the graph into two slightly overlapping parts promotes legibility, and thin identical rectangles identify the overlap. Compiled by author

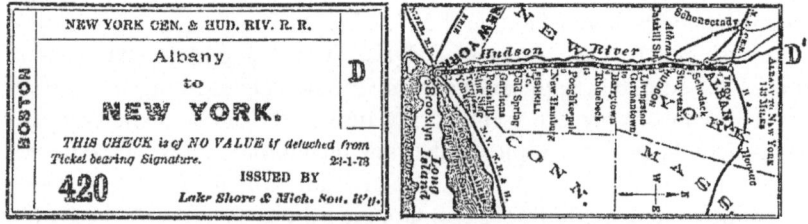

Fig. 1.5 One of the six examples used to illustrate Charles H. Waite's invention that added a map to the back of a railway ticket (US Patent 153,507; 1874)

end are two patents filed in 2009 that attest to the decline of the single inventor and the rise of electronic cartography. "Methods and Apparatus for Generating a Navigation Chart," filed by six inventors working for Boeing, was assigned to thirteen USPC categories, mostly focused on data processing and electronic display, while "Road Map with Indicated Road Segments," which involved three inventors and six categories, discusses paper maps as well as GPS displays.[31] Because a few patents issued after 1999 were filed seven years earlier, my graph is probably missing two or three valid patents filed before 2010.

The pronounced clumps between 1910 and the late 1920s and between 1980 and 2000, with a lackluster half century in between, is only partly explained by wartime distractions, economic downturns like the Great Depression of the 1930s, and the delayed impact of the Third Industrial Revolution, which began making non-digital technology obsolete after World War II.[32] Nor is it related to the general trend in patenting between 1880 and 2012, which I explored by calculating the statistical correlation between the cartographic patents in my database and patents in general. I restricted the analysis to utility patents because only 2 of the 304 patents were design patents, and I used award dates rather than filing dates, which were not readily available for patents in general.[33] The resulting correlation coefficient of 0.16 seems remarkably low, and using five-year running averages to dampen the year-to-year variations raised the estimate to only 0.26.[34] As a genre of creativity, patented cartographic inventions diverge markedly in frequency from the overall trend.

This divergence at least partly reflects my focus on printed maps. Wary that a low frequency of patenting in the post–World War II era might reflect an administrative foible, I asked the USPTO for an explanation. David Delzingaro, a writer-editor in the Classification Section, provided insights on the classification process and its history as well as the enlightening observation that "the relative dearth of printed maps patents issued from 1966–1977, including zero patents [issued] between '66 and '71, tends to confirm that printed mapmaking in terms of new technology was becoming a 'dying' art."[35] Although Delzingaro validates the oft-repeated prediction of the death of the printed map, a legacy of the disruptive emergence of digital cartography in the early 1970s, the critical phrases here are "printed mapmaking" and "new technology." Although mapmaking technology has moved well past the era when its intended product was a printed map, printed maps are now more abundant than ever, thanks to the proliferation of low-cost printers.

How then to account for the surge in map-related patents since 1980 (Fig. 1.4)? The answer largely reflects innovative folding methods that made printed maps more useful to consumers. I discovered the importance of folding innovations after reading and attempting to assign all 304 patents to subjective groupings that reflect both an invention's intent and the element of novelty.[36] Some categories, like "Map projection," were clearly distinct, akin to Supreme Court Justice Potter Stewart's take on hard-core pornography: "I know it when I see it."[37] Others, like "Travel or navigation aid," were megacategories that could be broken down further, for instance, to distinguish streetcar transfers from steam railway tickets, but seemed more meaningful when broadly construed. A few categories had cumbersome titles, like "Folding scheme predominant," which helped distinguish inventions focused on a novel folding method from inventions that include folding as an important but not overriding claim. I identified these latter patents by adding the attribute "Folding scheme mentioned," which could be checked off for patents that better fit another category. Although the 76 patents in the Travel category outranked the 60 patents in "Folding scheme predominant," addition of 32 patents in which folding was noteworthy but not dominant underscored the importance of convenient compactness.

Social scientists who study invention would not be surprised by the spurt of map folding patents in the 1980s and 1990s (Fig. 1.6). In 1922, political scientists William Ogburn and Dorothy Thomas, in a paper titled "Are Inventions Inevitable? A Note on Social Evolution," argued that multiple, seemingly independent inventions or discoveries are more the rule than an aberration.[38] They bolstered what's become known as the "Theory of Multiples" with a list of 148 breakthroughs made independently by two or more people. Their list includes advances in astronomy and celestial mechanics, biology, chemistry, electricity, mathematics, measurement and instrumentation, medicine, physics, and transportation and communica-

Patents filed, by year, 1874 to 2009, for maps (283/34) and indexed maps (283/35)

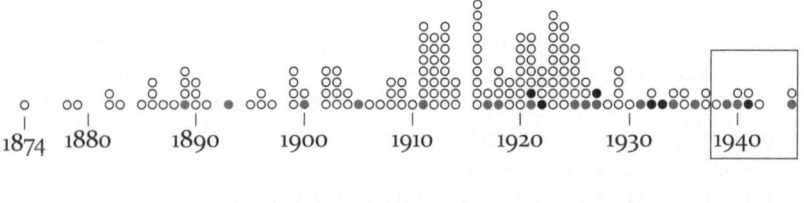

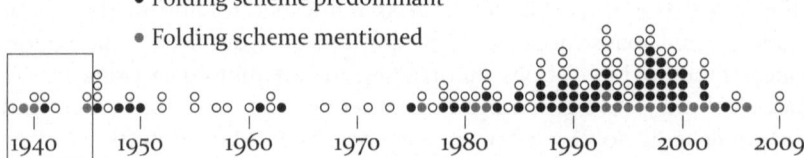

Fig. 1.6 Time-series plot highlighting patents concerned primarily (*black dots*) or incidentally (*gray dots*) with map folding. Compiled by author

tion. Some multiple discoveries occurred within a single year, as in 1747, when Alexis Clairaut, Leonhard Euler, and Jean d'Alembert devised independent solutions to the three-body problem involving the motions of the sun, the earth, and the moon. Others were more widely separated, as with the invention of the trolley car by Charles Van Depoele and Frank Sprague, in 1884–85 and 1888, respectively, following independent development of key elements in 1881 and 1883 by Ernst Werner von Siemens and Leo Daft, who experimented with electrically powered streetcars.

Ogburn and Thomas offered several complementary explanations, including the availability of essential constituents, the mental prowess of the inventor, and a variety of preconditions collectively labeled cultural preparation and easily understood insofar as "the problem has to be seen, its solution socially desired, and the ability must be trained and stimulated to attack the problem."[39] In this vein, the spurt of folded map patents in the late twentieth century, which I explore more fully in Chap. 4, no doubt reflects a combination of improved paper-folding machines that made these inventions practicable, a heightened concern for convenience among map users, and lively competition among producers of tourist and travel maps. That 48 different inventors or teams of inventors account for the 61 successful patent applications related to map folding filed in the United States between 1976 and 2006 suggests that making maps conveniently compact was recognized as a worthy challenge.

The simple streetcar transfer celebrated in this chapter exemplifies the synergy of essential constituents and cultural preparation. Although William Moffatt and Arnold von Schrenk were no doubt clever fellows, their invention is only partly a consequence of a mature trolley car network, which reflected prior discoveries and inventions such as the scientific understanding that electricity could be created, regulated, and transmitted and the subsequent development of electrical generators, transmission systems, and electric motors. By the 1880s steam railways had already demonstrated the efficiency of self-propelled vehicles running on iron rails. Decades earlier, horse cars had brought this technology to muddy city streets, and escalating urban growth in the latter half of the nineteenth century created a demand for improved public transit. The resulting network of carlines inspired clever ways for regulating trips that began on one route and ended on another.

Because the cultural preparation was right, it's not surprising that my database includes patent 1,161,312, for an application filed on 8 June 1911, and awarded on 23 November 1915 to William J. Hughes, of

Newark, New Jersey. Titled "Transfer-Ticket" like the patent issued the following year for Moffatt and von Schrenk's graphic map, Hughes's invention was based on a verbal map with only a linear list of transfer options (Fig. 1.7). His application also included a graphic map to explain the configuration of transfer points, but I doubt that the Albany inventors, who had filed in March 1913, knew of his work.

Although patent examiner J. Frederick MacNab might have known of the Hughes patent, it is not mentioned in his polite but increasingly contentious correspondence with the attorneys for Moffatt and von Schrenk, whose application was rejected three times, in April 1913, April 1914, and April 1915.[40] Each time Briesen & Knauth took almost a year to submit a markedly longer rebuttal—obviously the three years between filing and approval did not reflect a backlog in Washington. The first rejection chided the applicants for "failing to patentably distinguish from" Waite's 1874 patent (Fig. 1.5), "in view of the fact that it is notoriously old to punch data indicative of the general direction in which the passenger is traveling."[41] In a reply that included several strategic changes in wording, the attorneys argued that "the patent to Waite shows nothing but a partial map on the back face of a railroad ticket," and that the two applicants "are experienced street railway officials and are thoroughly familiar with the requirements thereof, and particularly with the transfer systems connected therewith."[42] Unconvinced—and apparently unaware that Moffatt was

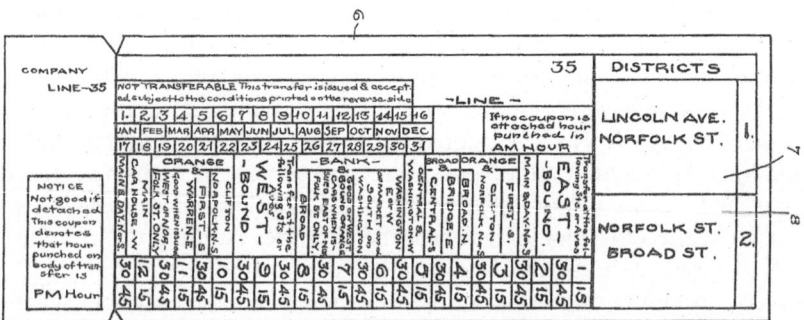

Fig. 1.7 Drawing of a streetcar transfer with a verbal map invented by William J. Hughes. In the patent application, this drawing (labeled Fig. 2, but with its long axis vertical) was juxtaposed with a graphic map (labeled Fig. 1) used to explain the transfer points. The *Official Gazette* for 23 November 1915 included only the drawing of the transfer, with its long axis horizontal, as in this view but much reduced (US Patent 1,161,312; 1915)

not actually a streetcar executive—MacNab responded with three short paragraphs, each rejecting two or more claims, and argued that "a mere punch mark through a diagrammatic representation of connecting railway routes would indicate at most that the transfer was issued by a conductor traveling in some direction thereon."[43] The attorneys, who responded with further rewording, again challenged the relevance of the Waite patent, while "trust[ing] that on re-examination the Examiner will admit the correctness of applicant's contentions and pass the case to an early allowance."[44] No doubt annoyed, MacNab repeated his point that printing a diagrammatic representation on the transfer was "old," and that "it makes no difference what kind of railway the tickets are used on." Tossing down a verbal gauntlet, he concluded, "Applicant may consider this rejection final for purposes of appeal if he so desires."[45]

As in many bureaucratic battles, persistence can be a winning strategy. A year later, in their third rebuttal, the attorneys—now Briesen & Schrenk—again asserted that punching the map to indicate direction was indeed a significant innovation and that references to Waite's patent and two others mentioned in the third rejection "do not affect the patentability of said claims."[46] MacNab and his boss might have grown tired of this sparring, or perhaps they concluded the Patent Office would lose if their wisdom were challenged in court. The next and last document in the file is a letter, dated 12 August 1916, and signed by Thomas Ewing, Commissioner of Patents, saying that the application "had been examined and ALLOWED."[47] The patent system is similar to the standard cartographic literature insofar as a stubbornly unwavering author can sometimes wear down a reluctant editor.

Nitpicking at the Patent Office clearly did not suppress the surge of cartographic patents in the 1910s and 1920s for inventions related to travel or navigation (Fig. 1.8). Cultural preparation is clearly evident in the increased ridership of steam and street railways as well as in vastly improved public roads and growing ownership of automobiles. Although many would-be patent holders were no doubt forestalled by competitors who had filed earlier or were discouraged by strict patent examiners, others had the right mix of cleverness and persistence to register promising ideas for planning sales trips, following routes on highways, or charting positions at sea. As subsequent chapters will demonstrate, the patent system's historical record offers intriguing insights to the motives of innovators as well as the winds of opportunity and waves of cleverness that prevailed outside the conventional currents of academic and government cartography.

Patents filed, by year, 1874 to 2009, for maps (283/34) and indexed maps (283/35)

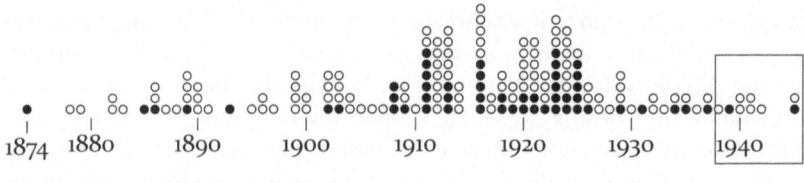

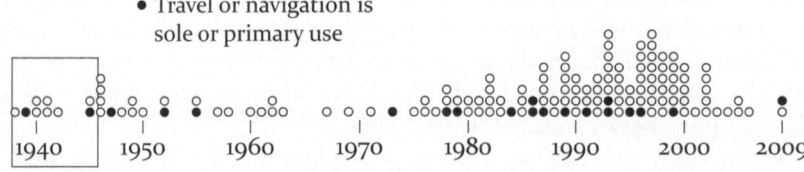

Fig. 1.8   Time-series plot highlighting patents concerned largely with travel or navigation. Compiled by author

NOTES

1. William C. Moffatt and Arnold von Schrenk, "Transfer-Ticket," US Patent 1,201,605, filed 3 March 1913, and issued 17 October 1916.
2. Streetcar transfers attracted the attention of William James Sidis, a brilliant but troubled scholar, who made street railway transfers a hobby and published a detailed, 306-page monograph titled *Notes on the Collection of Transfers* (Philadelphia: Dorance and Company, 1926) under the pseudonym Frank Folupa. For insights to Sidis's obsession, see Amy Wallace, *The Prodigy: A Biography of William James Sidis, America's Greatest Child Prodigy* (New York: Dutton Adult, 1986), esp. 181–99, 244–50. Dorance was a vanity publisher, and the book became rare (and expensive) after most of the stock was lost in a warehouse fire. An electronic copy is available online at http://www.sidis.net/TransfersContents.htm.
3. In 1995, the term of a utility patent increased from 17 years after the issue date to 20 years after the filing date. For discussion of the term of patent protection, see Ben Ikenson, *Ingenious Inventions: How they Work and How They Came to Be* (New York: Back Dog & Leventhal Publishers, 2004), 274–75; and Mark Lemley, "An Empirical Study of the Twenty-year Patent Term," *AIPLA Quarterly Journal* 22 (1994): 369–424.

4. I say late 20s because the Social Security Death Index reports his birthdate as 22 June 1883, which would have made him 29, whereas a manuscript coding sheet for the 1930 Census lists him as 42 years old, which would have made him about 25.

5. Entries for Moffatt in city directories in Albany and Scranton were found using the Library edition of Ancestry.com, as were several family trees and listings in federal census manuscripts and the Social Security Death Index. Online searching and Newspaper.com led to articles in the *Scranton Republican* and the *Hazleton (PA) Plain Speaker*.

6. "Hundreds Witness Passing of Trolley," *Scranton Republican*, 7 August 1934, 5.

7. The principal source of background information on von Schrenk was Ancestry.com, which was a portal to his listings in several city directories as well as a New York University alumni directory, his World War I military draft and service records, and his North Carolina death certificate. *The Electric Railway Journal* for 23 January 1915 (vol. 45, no. 4, p. 206) reported that he had left his job as general superintendent of United Traction in Albany to work for his brother's firm of consulting engineers in St. Louis.

8. The ghostwriter analogy is from Kara Swanson, "Authoring an Invention: Patent Production in the Nineteenth-Century United States," in *Making and Unmaking Intellectual Property: Creative Production in Legal and Cultural Perspective*, ed. Mario Biagioli, Peter Jaszi, and Martha Woodmansee, 41–54 (Chicago: University of Chicago Press, 2011), esp. 45–49.

9. I found no mention of maps on the fronts of transfers in William Sidis's *Notes on the Collection of Transfers*, a remarkably thorough book readily dismissed as supremely boring by readers who don't share its author's hobby or appreciate the meticulousness of an at least mildly obsessive collector. Sidis discussed various non-cartographic schemes for representing transfer options and used maps to inventory street railway systems in the northeast.

10. Fame and recognition are often as important as, if not more important than, power, money, and security as drivers of innovation. Devrim Goktepe and Prashanth Mahagaonkar, "What Do Scientists Want: Money or Fame?" *Jena Economic Research Papers* no. 32 (2008) [online at econpapers.repec.org/paper/jrpjrpwrp/]; and

Michael A. Gollin, *Driving Innovation: Intellectual Property Strategies for a Dynamic World* (New York: Cambridge University Press, 2008).

11. Juan Alcácer, Michelle Gittelman, and Bhaven Sampat, "Applicant and Examiner Citations in U.S. Patents: An Overview and Analysis," *Research Policy* 38 (2009): 415–27.

12. Article I, Section 8 of the Constitution lists among the prerogatives of Congress the power "To promote the Progress of Science and useful Arts, by securing for limited Times to Authors and Inventors the exclusive Right to their respective Writings and Discoveries."

13. For discussion of how the patent system encourages inventors to build on prior innovation, see James Bessen and Eric Maskin, "Sequential Innovation, Patents, and Imitation," *RAND Journal of Economics* 40 (2009): 611–35.

14. For a history of the Patent Office, see Kenneth W. Dobyns, *The Patent Office Pony: A History of the Early Patent Office* (Fredericksburg, VA: Sergeant Kirkland's Museum and Historical Society, 1997).

15. William C. Moffatt and Arnold von Schrenk, "Transfer-Ticket," *Official Gazette of the United States Patent Office* 231 (17 October 1916): 738–39.

16. Gustavus Adolphus Weber, *The Patent Office: Its History, Activities and Organization* (Baltimore: John Hopkins Press, 1924), 49. The Institute for Government Research merged with the Brookings Institution in 1927.

17. William J. Rankin, "The 'Person Skilled in the Art' Is Really Quite Conventional: U.S. Patent Drawings and the Persona of the Inventor, 1870–2005," in *Making and Unmaking Intellectual Property: Creative Production in Legal and Cultural Perspective*, ed. Mario Biagioli, Peter Jaszi, and Martha Woodmansee, 55–75 (Chicago: University of Chicago Press, 2011); quotation on 66.

18. The extended term of copyright protection largely reflects aggressive lobbying by Walt Disney Productions, horrified that Mickey Mouse might pass into the public domain. Among the numerous legal essays on what might become a perpetual copyright, see William M. Landes and Richard A. Posner, "Indefinitely Renewable Copyright," *University of Chicago Law Review* 70 (2003): 471–518.

19. George Cho, "Intellectual Property," in *Cartography in the Twentieth Century* (Vol. 6 of *The History of Cartography*), ed. Mark Monmonier, 654–59 (Chicago: University of Chicago Press, 2015).

20. Since 1947 the Patent Office has required citations of earlier patents as prior art in much the same way that contributors to the standard literature cite previously published articles, maps, and monographs. See Alcácer, Gittelman, and Sampat, "Applicant and Examiner Citations in U.S. Patents: An Overview and Analysis."

21. Weber, *The Patent Office*, 19.

22. Weber, 65–66, 89; and Fred K. Carr, *Patents Handbook: A Guide for Inventors and Researchers to Searching Patent Documents and Preparing and Making an Application* (Jefferson, NC: McFarland, 1995), 129.

23. "Overview of the U.S. Patent Classification System (USPC)," December 2012, page I-3. Online at www.uspto.gov/sites/default/files/patents/resources/classification.overview.pdf.

24. Quotation from David Fitzpatrick, USPTO Office of Patent Planning and Capacity Analysis, email, 18 February 2015. Also see "USPTO and EPO Announce Launch of Cooperative Patent Classification System," USPTO press release 13-01, 2 January 2013.

25. *Manual of Classification of Subjects of Intention of the United States Patent Office, Revised to January 1, 1916* (Washington, DC: Government Printing Office, 1916), 30. Also see note 15.

26. The current classification schedule is online at http://www.uspto.gov/web/patents/classification/index.htm (accessed 17 February 2015).

27. For discussion of commonly understood definitions of *map* and *cartography*, see Christopher Board, "Map: Definitions of Map," in *Cartography in the Twentieth Century* (Vol. 6 of *The History of Cartography*), ed. Mark Monmonier, 798–801 (Chicago: University of Chicago Press, 2015).

28. For discussion of Google Patents, see Andrea L. Hamilton, "Putting Google Scholar to the Test on Patent Research," *Colorado Lawyer* 39.5 (2010): 79–81; David Hitchcock, *Patent Searching Made Easy*, 5th ed. (Berkeley, CA: Nolo, 2009), 137–50; and John J. Meier and Thomas W. Conkling, "Google Scholar's Coverage of

the Engineering Literature: An Empirical Study," *Journal of Academic Librarianship* 34 (2009): 196–201.

29. For historical analysis, the date of application is more relevant than the date of issue because "in most cases, the inventive activity reflected by a patent application *began* about a year or two before the application was filed." Jacob Schmookler, *Invention and Economic Growth* (Cambridge, MA: Harvard University Press, 1966), 22.

30. Charles H. Waite, "Improvement in Tickets," US Patent 153,507, filed 14 April 1874, and issued 28 July 1874.

31. These patents, numbered 8,260,545 and 8,094,043, were issued on 4 September 2012, and 10 January 2012, respectively.

32. Jeremy Greenwood, "The Third Industrial Revolution: Technology, Productivity, and Income Equality," *Economic Review* 35.2 (1999): 2–12.

33. Yearly counts for all utility patents by year of issue are from US Patent and Trademark Office, "U.S. Patent Activity, Calendar Years 1790 to the Present," http://www.uspto.gov/web/offices/ ac/ido/oeip/taf/h_counts.htm. Counts tabulated by year of filing were not available. Corresponding counts for cartographic patents were tabulated by the author.

34. Squaring the correlation coefficient yields the coefficient of determination, which ranges from 0 to 1, and in this context measures how well one time series matches another. Statistically, this means that the general temporal trend in patenting accounts for less than seven percent of the temporal variation in map-related patents. Assuming four-decimal-place accuracy, the coefficients of correlation and determination based on five-year averaging for 1882 to 2010 were 0.2562 and 0.0656, respectively. Corresponding estimates calculated without averaging for 1880 to 2012 were 0.1626 and 0.0264. When I tried fitting a straight line to the yearly counts in Fig. 1.4, the result was a barely noticeable upward slope, reflected by a correlation coefficient of only 0.06 and not worth plotting. By contrast, utility patents for all inventions yielded a much stronger (0.81) linear correlation with time. Repeating these analyses with five-year averaging produced modest increases in both correlation coefficients, to 0.09 and 0.83, respectively—further evidence that the patenting of map-related inventions was out of step with the patenting of inventions in general.

35. David Delzingaro, email, 23 August 2013.
36. My reading of individual patents revealed that the Patent Office had apparently incorrectly assigned several patents to categories 283/34 or 283/35. A few of these patents (e.g., Patent 3,744,541, issued 10 July 1973, to Harold Brauhut, of Honey Toy Industries, in New York City, for an invention titled merely "Wallet") were blatantly non-cartographic—perhaps the result of classification or transcription errors at the USPTO—while others might have been cartographically relevant had they included a map in their drawings or mentioned maps specifically in their text. Their presence underscores the need to question the USPTO's implementation of its classification scheme.
37. See Stewart's oft-quoted concurring opinion in *Jacobellis v. Ohio*, 378 U.S. 184,197 (1964).
38. William F. Ogburn and Dorothy Thomas, "Are Inventions Inevitable? A Note on Social Evolution," *Political Science Quarterly* 37 (1922): 83–98. Technology journalist Clive Thompson observed that "the discussion of multiples is, as you might expect, itself a multiple." See his *Smarter Than You Think: How Technology Is Changing Our Minds for the Better* (New York: Penguin Press, 2013), 299n58.
39. Ogburn and Thomas, 92.
40. Correspondence related to individual patents is organized by patent number as part of Record Group 341.3, stored at the National Archives site in Kansas City, Missouri.
41. J. F. MacNab (examiner) to Briesen and Knauth [sic], 30 April 1913.
42. Briesen & Knauth to Commissioner of Patents, 14 February 1914.
43. J. F. MacNab (examiner) to Briesen & Knauth, 6 April 1914.
44. Briesen & Knauth to Commissioner of Patents, 8 March 1915.
45. J. F. MacNab (examiner) to Briesen and Knauth [sic], 22 April 1915. Applicants displeased with a Patent Office decision could appeal to the Commissioner of Patents, and if not satisfied, could appeal the Commissioner's decision to the US Court of Appeals for the District of Columbia. Weber, *The Patent Office*, 32, 38–39.
46. Briesen & Schrenk to Commissioner of Patents, 5 April 1916.
47. Thomas Ewing (Commissioner of Patents) to Wm. C. Moffatt and Arnold von Schrenk, 12 August 1916.

CHAPTER 2

Pinpointing Places

Although academic cartographers prefer to think of maps as tools for understanding our world, cartography's most basic need is an addressing system that assigns each place an unambiguous location. Latitude and longitude work well for a perfectly spherical earth—assuming everyone agrees on the Prime Meridian—but crustal geophysics and cruise missiles require more exact geometries that accommodate the spinning planet's flattening at the poles and other geodetic anomalies. At a more quotidian level, letter carriers and delivery drivers appreciate named streets and numbered houses, and atlas users know that page numbers combined with grid cells arranged in numbered rows and lettered columns are sufficient for finding places listed in the index. However well established, these old standbys for indexing location were not immune to patented improvements by clever innovators.

In focusing on georeferencing, this chapter begins with a time-series analysis similar to those in Chap. 1, for folding schemes and travel. And as with map folding, the time-series graph for georeferencing (Fig. 2.1) requires two levels of intensity, with black dots denoting patents devoted largely to referencing location and gray dots denoting patents more logically assigned to another megacategory even though location is a key element. The resulting graph shows location indexing concentrated in the first two decades of the twentieth century and peaking well before travel mapping (Fig. 1.8) and map folding (Fig. 1.6).

© Mark Monmonier 2017
Mark Monmonier, *Patents and Cartographic Inventions*,
Palgrave Studies in the History of Science and Technology,
DOI 10.1007/978-3-319-51040-8_2

Patents filed, by year, 1874 to 2009, for maps (283/34) and indexed maps (283/35)

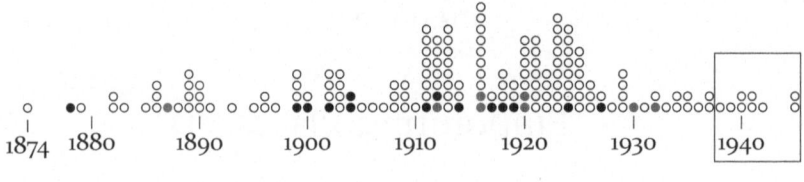

• Georeferencing scheme predominant

• Georeferencing scheme mentioned

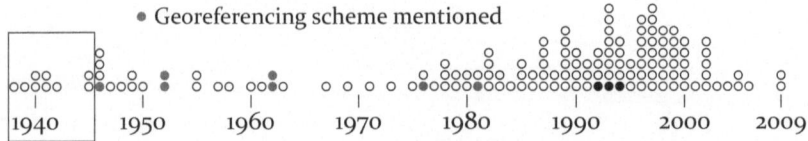

Fig. 2.1 Time-series plot highlighting patents concerned importantly if not predominantly with geographic referencing. Open dots represent other patents in the dataset. Compiled by author

This dearth of more recent patents is telling. By contrast, ingenious late nineteenth century innovations in georeferencing held not the slightest hint of a dying art. The earliest black-dot patent was filed in 1880 by Chicago resident Samuel Gross, who proposed printing maps on a "flexible material" with ruler-like scales along the edge on the reverse side so that the distance between a pair of points could be estimated by folding over the map to align one of these scales with the two points.[1] Although Gross's drawing (Fig. 2.2) too conveniently emphasizes distances parallel to the map's edges, folding back the map works well for other orientations. It's a clever, why-didn't-I-think-of-that idea that correctly assumed the availability of both a durable flexible material and a suitable printing method: prerequisite technologies that might have inspired his invention.[2]

The earliest patent in my gray-dot subcategory was filed in 1887 by William A. Baugh, a Melbourne, Florida, resident who anticipated his state's land boom by four decades with an invention titled "Methods of Subdividing and Designating Land."[3] Better suited to the megacategory Land Records (too skimpy for this book), his invention is a nested hierarchy of rectangular grids aptly described by the first four of his five drawings (Fig. 2.3). Although land surveyors and historians of the American West will quickly recognize the meandering sequence of the 36 numbers

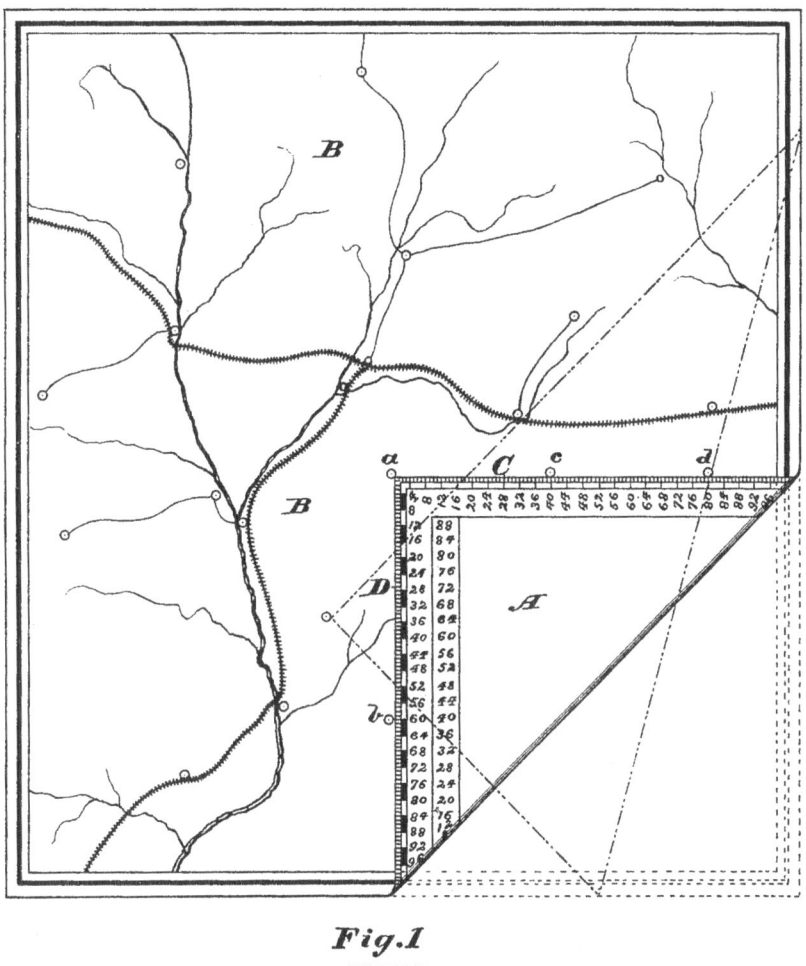

Fig. 2.2 Map printed on flexible paper with distance scales on the back allows the user to measure direct distances between two points, for example, between *a* and *d*, which are 80 units apart, as shown (US Patent 232,261; 1880)

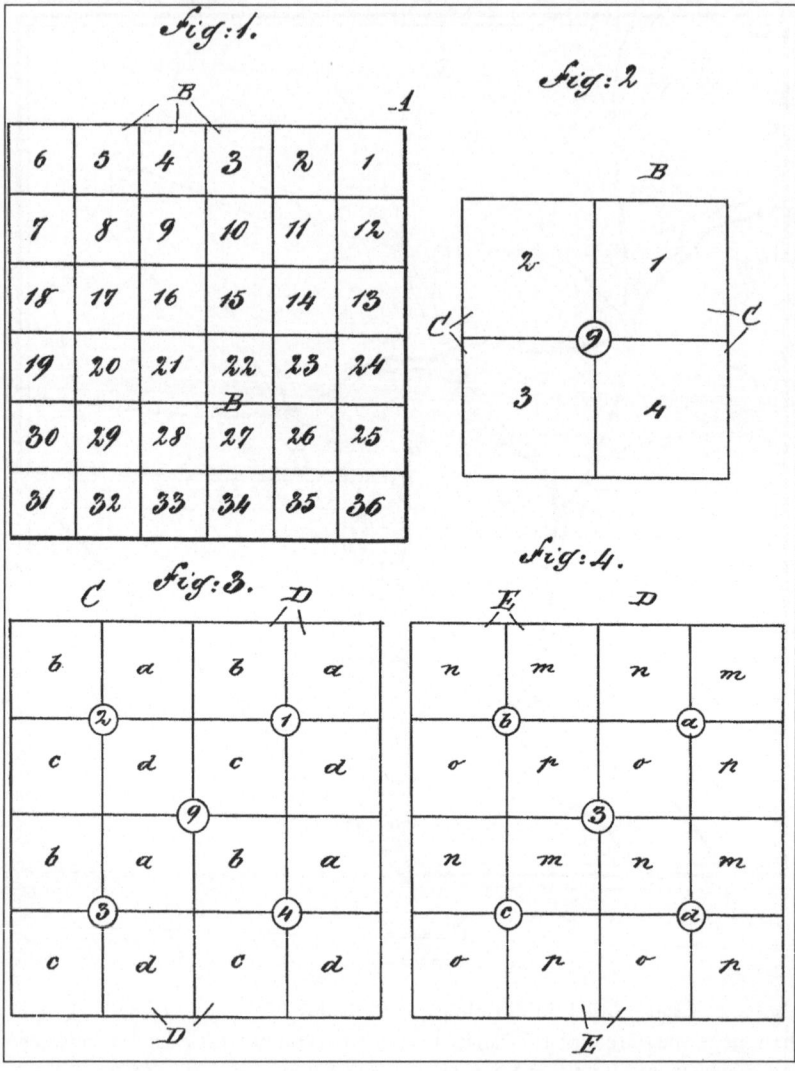

Fig. 2.3 Drawings describing the first three successive subdivisions of the 36 square-mile sections of a township in the US Public Land Survey System (*upper left*). Sections one mile on a side and containing 640 acres (labeled B in the *upper left*) are each subdivided into four quarter-sections of 160 acres (labeled C in the *upper right*), each subdivided in turn into four 40-acre square lots (labeled D in the *lower left*), and again into 10-acre square lots (labeled E at the *lower right*) (US Patent 367,178; 1887)

in the 6 × 6 grid at the upper left as the numbering scheme for square-mile sections in the US Public Land Survey System, digital cartographers will identify Baugh's nested hierarchy of 2 × 2 "quad" subdivisions as a precursor of the "quadtree" data structure that became a staple of academic digital cartography in the 1970s.[4] Successive division of a square section into quarters, numbered 1–4 in the first round and lettered *a–d* in the second round and *m–p* in the third, narrows the reference down to a square 1/64 (¼ × ¼ × ¼) the size of the original; in this way any of these 64 small squares can be referenced by a sequence of one integer and two letters. That late-twentieth-century academic cartographers were unaware of Baugh's invention underscores my point about mutually unaware parallel literatures.

The most recent black-dot invention in the time series (Fig. 2.1) is a patent filed in 1994 by John A. Jones of Alvin, Texas, who had devised "an alphanumeric system ... for easily and quickly locating and identifying a particular section of a geographical region which may range in size from several miles square to several feet square."[5] His specific claims include an assertion that the system would be particularly useful "in rural areas which do not have a conventional street numbering system."

Conceptually similar to Baugh's hierarchical four-cell, 2 × 2 grid structure, Jones's invention called for a hierarchical nine-cell, 3 × 3 grid structure with the numbers 1 through 9 identifying the cell retained after each successive subdivision. His simple graphic narrative (Fig. 2.4) proceeds upward from hypothetical area J (bottom layer), divided initially into nine cells. Because cell 3 (upper right in the 3 × 3 grid) remains after the first split, the next layer up is called J3. And because cell 2 (top center in the 3 × 3 grid) survives the second split, the third layer up is J32. Subsequent splits at cells numbered 6, 7, 5, 8, and 4 reference an area called J3267584 that is only 1/4,782,969 (1/9⁷) the size of area J. If J is a rectangular area of 900 square miles—a square 30 miles on a side, say—seven successive splits would precisely locate an area less than 1/8 acre, roughly the size of small residential lot. Another level of splitting would further shrink the area to another ninth.

Although Jones's "Geographic Location Identification System" crammed an impressively exact location into a simple string of integers, he apparently was unsuccessful in exploiting his patent. According to the website Patent Buddy, the patent expired in 2003 after he failed to pay the mandatory maintenance fee.[6] In 1980 Congress had revised the patent law to include an escalating "annuity fee" to be paid 3½, 7½, and 11½ years

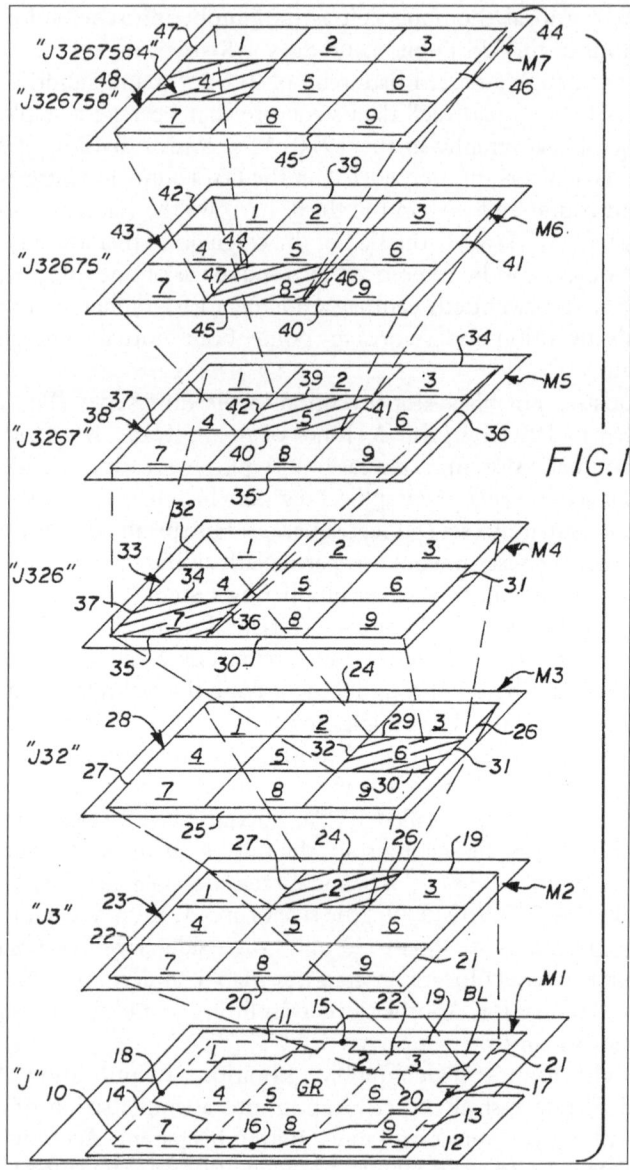

Fig. 2.4 Graphic explanation of John Jones's hierarchal strategy for pinpointing a location through a series of successive divisions of an area with a nine-cell 3 × 3 grid. After each nine-way split, a single digit is added at the right end to indicate the cell remaining (US Patent 5,445,524; 1995)

after the award date in order to keep a utility patent in force.[7] One goal was to increase revenue so that the patent system could remain self-supporting; another was to discourage rights holders from maintaining patents that were not being used—dormant patents, often filed prematurely to forestall competitors, are believed to stifle innovation.[8] The fee jumped significantly between the first and second payments. In 2003 the 7½-year maintenance fee would have been more than $2000, which Jones apparently decided wasn't worth it.[9] He was not exploiting the invention himself, and licensing was unlikely because quadtree schemes based on four pairs of binary digits (00, 01, 10, 11) were not only more computationally efficient but also well known in the academic literature, and presumably were in the public domain.

If commercial exploitation of a patent is the criterion for success, none of the georeferencing inventions in my dataset is as impressive as John Byron Plato's addressing and mapping scheme that inspired a series of "rural directories" published for western and central New York State counties in the 1920s and 1930s. Plato's patent, ambiguously titled "Map or Chart," was filed in early December 1914 and approved within eight months.[10] Although he took a few years to find the right backers, by the end of the decade he had moved to Ithaca, New York, where local businessmen and officials at the New York State College of Agriculture, a division of Cornell University, helped him set up a firm that grew to serve thousands of farmers and hundreds of related businesses and provided him employment through the remaining life of the patent. Ironically, his map supplied a straightforward answer to the rural address problem John Jones claimed to solve eight decades later.

Plato based his invention on the notion that an obscure rural destination could be described succinctly, at least at first cut, by its distance and direction from a town or village with a post office, one or more schools and churches, and various businesses serving the surrounding area. Because "five miles due east of Centreville" is too vague for pinpointing rural destinations, he recognized the need for a map that not only subdivided the area surrounding the central place into zones but also assigned individual residences within each zone a unique identifier, much like the house numbers along a city street. In awkwardly precise patentese, he proposed "a system of designating dwellings in rural districts, by numbers or other distinctive symbols, principally for the purpose of facilitating the delivery of mail by what is commonly known as the rural free delivery service." Classic patent-speak.

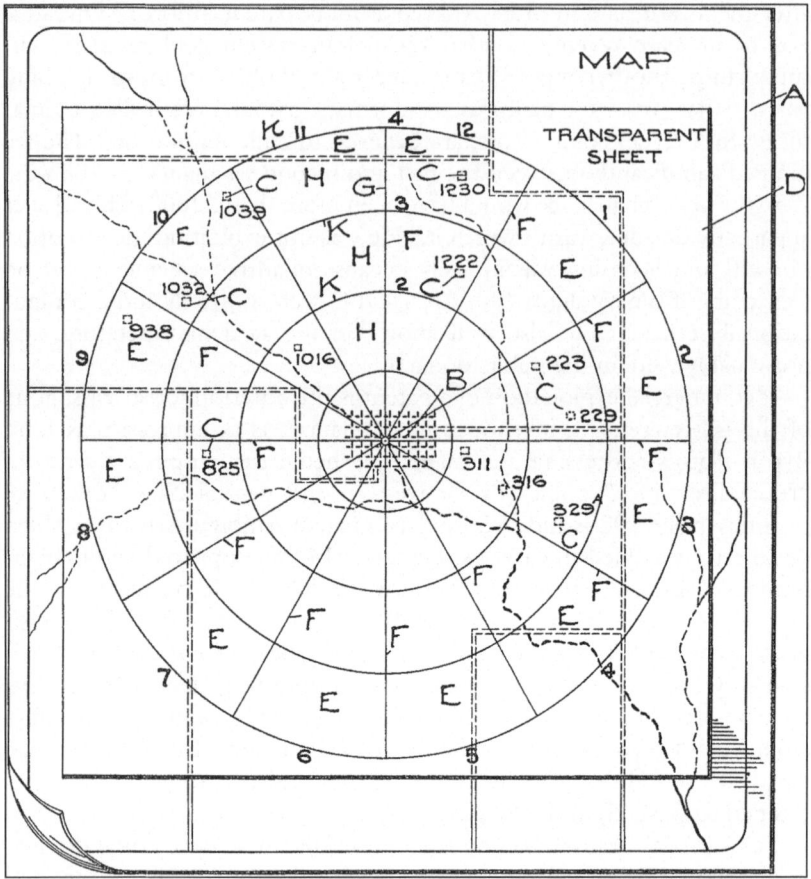

Fig. 2.5 Basic framework of John Byron Plato's Clock System for rural addresses (US Patent 1,147,749; 1915)

Instead of dividing the area surrounding a central "distribution point" into the eight octants of the compass card (N, NE, E, etc.), Plato sliced it into twelve 30-degree sectors "numbered in clockwise succession from 1 to 12." Concentric circles a mile apart then subdivided each sector into zones, and within each zone, unique numbers or symbols pinpointed specific dwellings. As shown on his patent's single drawing (Fig. 2.5), these numbers could also reference both the sector and the ring. For example,

1039 (toward the upper left in his drawing) refers to the 9th farmstead in a zone in the 10 o'clock sector between the 3- and 4-mile circles. To accommodate more than ten destinations within a zone, the patent's multiple claims allowed labeling schemes based on "different characters," including letters of the alphabet, used in later implementations.

The label "transparent sheet" near the upper right in the drawing does not reflect Plato's ultimate development of his patent. Although the Clock System could be implemented as a portfolio of graphic templates, one for each community center in the region, it was more efficient to partition the region into service areas, each with its own clocklike framework, and to plot all frameworks and destinations on a single map. Individual dwellings would then have a unique address consisting of the name of the central place, the Clock System coordinates, and a point identifier, and an index accompanying the map could list all dwellings alphabetically by the resident's last name. Plato eventually realized that making and marketing maps and directories was a better business plan than trying to license his invention to the Post Office.

That the Patent Office approved his application in less than eight months reflects a mix of creativity, self-confidence, and willpower sufficient to offset an astonishing naiveté. Plato's case file in the National Archives includes a cover letter, a specification and list of claims, and an oath affirming that he was a US citizen, a resident of Semper, Colorado (now a suburb of Denver), and "the original, first and sole inventor of the improvements in map designing and mail directing systems" described therein—all written out longhand.[11]

Six weeks later, the Patent Office sent a rejection letter citing various shortcomings. Plato's description of the invention did not meet "the standard required by the office" and "lack[ed] the customary preamble."[12] His drawing "was received in a mutilated condition," acceptable for examination but not suitable for publishing. More damning, his claims referred to a "so-called system [that] involves merely the idea, rather than a means for carrying out the invention." Nonetheless, a remedy was apparent: because his drawing "illustrated a map or chart for carrying out the system, it is to this chart that claims should be drawn." And after rejecting the claims because "they do not cover patentable subject matter," the examiner, L. D. Underwood, also acknowledged "the difficult nature of the subject matter of this application and the difficulty presented in attempting to properly and adequately cover same in a claim." The applicant, he suggested, "would best serve his own interests by seeking the aid of competent, registered patent counsel." Plato took the advice, and hired attorney G. J. Rollandet (G for Gerrit, J for Jan), with whom he had worked before.

Rollandet was more than just a patent attorney. His half-page ad in the 1915 Denver city directory (with "Registered Solicitor of PATENTS" in bold type) supplemented smaller listings under multiple headings: Blueprinting, Drawings—Foreign and Caveats, Draughtsmen, Map Designers, Map Mounting, Map Publishing, Patent Attorneys and Solicitors, Patent Office Drawings, Patents, and Patents Developed.[13] Rollandet, who was born in The Netherlands in 1866 and emigrated to the United States in 1888, was not shy about advertising. He had secured— and no doubt paid for—a page in *Representative Men of Colorado in the Nineteenth Century*, a 272-page photographic "portrait gallery" pub- lished in 1902, which identified him as "Manager, Rollandet Blueprint and Drafting Company."[14] The accompanying portrait shows a serious- looking young man with a full black moustache, pince-nez glasses, and hair parted in the middle. Lawyering in the early twentieth century was not credentials conscious: a university degree was not required and asser- tive self-promotion was not frowned upon. In the 1935 city directory, his listing under Trade Marks proudly proclaimed "36 years experience [with] U.S. and Foreign Patents & Trade Marks."

In less than six weeks, Plato had signed a power of attorney, which Rollandet mailed to the Patent Office along with a new drawing, a revised specification, a reworked set of claims, and a request to rename the invention "Map or Chart"—a phrase prominent in Underwood's rejection letter.[15] Five weeks later, Underwood rejected all 17 claims, mostly because of phrasing apparently preempted by four existing pat- ents.[16] For example, claims 1 through 8 were rejected because of simi- larities to Baugh's quadtree scheme for subdividing land (Fig. 2.3) and to a patent recently awarded J. H. Robinson for a "Tariff Chart" that fit state boundaries to a grid of equal-size square tiles with unique numbers useful for specifying shipping rates (Fig. 2.6).[17] Although neither pat- ent's drawings bears much resemblance to Plato's, there were trivial and inescapable similarities, which the patent examiner cautiously pointed out. "While the Baugh patent does not show the distinctive designating symbols as used to show the various land subdivisions," he wrote, "it would not involve invention to place such designations on the map in view of the fact that Robinson shows a map with the various zone desig- nations appearing thereon." Had the examiner crossed the line between close scrutiny and nitpicking?

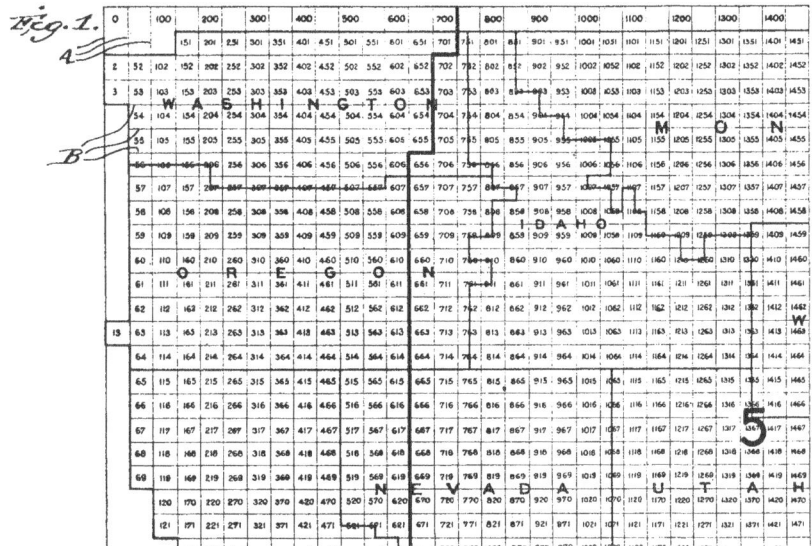

Fig. 2.6 Excerpt from sheet 1 of J. H. Robinson's "Tariff Chart." Thin lines divide the country into rectangular areas with numbers that refer to the rows in a rate chart; medium-weight lines are state boundaries, and bold lines delineate "charge zones" with a specific set of shipping rates (US Patent 1,029,085)

Rollandet replied five weeks later, after having changed several words in the specification and substantially reworked the claims.[18] Claim 15, after some strategic rewording, was now claim 1, and new claims numbered 2 through 8 replaced the other 16 claims. Underwood, who replied two and a half weeks later, pronounced the claims "allowable, as at present revised."[19] Most of his short letter explained the need for one further change in the specification, from "districts [plural] to the left" to "district [singular] to the right." He also noted that the substitute drawing sent earlier was acceptable. Rollandet quickly confirmed the revision and requested "early and favorable action."[20] Two weeks later the Commissioner of Patents informed Plato, through his attorney, that his patent had been "allowed."[21] Once the Patent Office received the "final fee" of $20, "the printing, photolithographic, and engrossing of the several patent parts, preparatory to final signing and sealing," could begin. Final processing would "require about four weeks," an accurate estimate insofar as Rollandet mailed the payment on June 22 and the patent was published on July 27.[22]

I am puzzled that Plato had originally filed the patent himself. He had received three patents since 1905 and should have known how the process worked, but he either overestimated his understanding of the lingo or thought attorneys were too expensive or too slow.[23] Rollandet had represented him for two of the three patents: one that took less than eight months and another that took more than five years. To Plato's credit, once he knew he needed an experienced attorney, he got one.

Plato's three patents describe a device for preventing a horse from running away with a wagon when the driver stepped away. A strap connected to a bit in the horse's mouth was also connected to a drum under the wagon. Before stepping down, the driver would throw a lever that engaged the drum to a wheel so that if the horse moved forward, the strap would wind around the drum and pull on the bit, thereby exerting pressure on the horse's gum; if not in panic, the horse would stop. When the horse backed up, the drum unwound automatically and the pressure stopped. This "Hitching Attachment for Vehicles" is functionally akin to a car's parking brake: a set-it-and-forget-it device enhanced by subsequent patents titled "Horse-Hitching Device" and "Hitching Device."

These were not vanity patents. Denver city directories published from 1906 through 1909 list Plato twice: once at the home he shared with his mother and again for the "Plato Manufacturing Co." (in large bold type), identified as "mfrs Star Horse Hitch, J. B. Plato mgr." Addresses two miles apart indicate that the horse-hitching firm was not a backyard or basement sideline.[24]

Genealogical research tools provide key details of Plato's life.[25] He was born in Chicago on 17 December 1876. According to manuscript records for the 1880 Census, at age three he was living in Geneva, Illinois, about 30 miles west of downtown Chicago, with his parents John Byron and Helen Plato, and his paternal grandmother. John senior, 37 years old, was a "retired lawyer" suffering from "consumption," later known as tuberculosis; he died in 1881. His son, the inventor, was an only child. Records from the 1890 Census were lost in a 1921 fire. City directories for 1898 and 1899 list him as a student living in Denver with his mother. Coding forms for the 1900 Census show Plato in Denver, sharing the home of his maternal grandmother with his mother and four other relatives; records list his occupation as draftsman but do not name an employer.

A military service question on the 1930 Census, identified Plato as a veteran of the Spanish-American War, fought in 1898–99, when he would have been 21 years old. I found him in a history of the First Colorado

Infantry, of the United States Volunteers, which was sent to fight the Spaniards in the Philippines.[26] He was a private, previously a student in Denver, and a member of Company E, 1st Battalion, which saw action in spring 1899. Seven soldiers in his unit of slightly more than 100 men were wounded in battle, one fatally. Plato returned with the others in 1899.

I found no evidence that he ever attended college but for the 1940 Census he reported having completed four years of high school. City directories for 1901, 1903, and 1904 report him in the lumber business, apparently self-employed and specializing in cabinet woods and veneers. Selling lumber was a convenient fallback after he left the horse-hitch business: the 1910 and 1911 directories list him as manager of the South Side Lumber Co., rounding out a skill set that included experience in drafting, inventing and patenting, and running a business. In 1912 he was a schoolteacher in a Denver suburb—perhaps teaching shop—and still living with his mother and grandmother. Further occupational playfulness was apparent in 1914, when the city directory listed him as a rancher and at a different address, this time without his mother. The 1916 directory reported that he had moved again but listed no occupation.

Other sources confirm that Plato, now nearly 30 years old, had become a livestock and dairy farmer. He also had at least a passing interest in writing, as demonstrated in the July 1917 issue of the magazine *System on the Farm*, which combined two short articles under the title "Reducing Barn Costs—Plans Used by Two Dairyman."[27] Plato was listed second in a byline followed by the teaser, "One has 30 cows, the other 11." His half, subtitled "A Plan for a Smaller Herd," described an 80 × 16 foot shed he had built himself using materials purchased for $60, a tenth the cost of a conventional barn. Open on one side, it contained a long feed rack and four stanchions, which confined the cows during milking. Having only four stanchions, instead of one for each animal, saved space, and "the time required to let the four cows out of the stanchions, and the other four in, amounts to almost nothing." And because winters in his part of Colorado were suitably mild, the open side was covered only by a canvas, which kept the cold out and body heat in on cold days and could be rolled up on warm days. "The cost of housing a cow is cut ninety per cent," Plato argued, and "with plenty of straw the cows are as clean as kittens." Eight additional advantages underscore his concern for the animals' health and comfort as well as for operating costs and the efficient use of space.

Plato's name also appeared occasionally in the *Guernsey Breeders' Journal*, which reported the buying and selling of these distinctive white-and-orange purebred dairy cattle. For example, the transfer list for 1–14 March 1918 noted that J. B. Plato, of Broomfield, Colorado, had acquired Flossie of the Bar Forks 39548 from N. M. Hubbard: 39548 was Flossie's registration number, and Bar Forks, probably the name of the farm where she had been born, was added to distinguish her from countless other Flossies.[28] The same list reported that Plato had sold a bull, Primus of Mar-Lence 48926, to C. W. Lothrop, of Arvada, Colorado.

Although the rural livestock trade relied on newspapers, agricultural journals, and the postal system to advertise availability and arrange deals, a buyer eager to inspect an animal often had to travel over poorly marked roads to an unfamiliar destination. This challenge inspired the invention of the Clock System, according to an interview with Plato in the April 1917 issue of *Illustrated World*, a popular magazine focusing on science and technology.[29] Some "eastern buyers" he had lined up stopped at the local post office to ask for directions to his farm, which they knew only as Box 41, R.F.D. 1, Bloomfield, Colorado. The clerk, who didn't know, suggested they ask the carrier, who would not return until the following morning. The frustrated buyers wrote Plato a polite note and left on the afternoon train. "That killed a mighty profitable bargain," he told the reporter, but "got me thinking about rural addresses—if you can call them that." What was needed, he reasoned, was "new ways of numbering farms." Maps could be helpful, but "if the problem was to be solved by mapping, as it is done by the experts, the big map houses would have had the answer long ago." What was needed, he realized, was "an entirely new element of location." While pondering possibilities, he was haunted by the phrase an acquaintance had uttered: "It takes time to find your house." The penny dropped the day his watch stopped. "As I stood scowling into its face it smiled back the answer." Let the familiar clock face provide direction from the village post office, and then add distance.

At the time of the interview, Plato apparently believed the Post Office would welcome his solution. According to the article, at least one postal official "told him that he had exactly what the Department had been looking for, and that they would be glad to incorporate it in their system." Perhaps so, but the Clock System was no match for bureaucratic inertia. A changeover would have incurred a significant start-up cost, the existing system that relied on the rural letter carrier's local knowledge worked fine, and the Post Office had no obligation to help travelers. The postal service continued to require RFD numbers, and Plato's "Rural Index" maps, published after 1920, cautioned against using the Clock System for addressing mail.

Although Plato might have told *Illustrated World* his addressing scheme "would be applicable without the use of a map or chart of any kind," he surely understood the map's multiple roles in explaining and promoting the system, assigning addresses, and providing a geographic framework. Indeed, a year earlier *The Daily Ardmoreite*, a southern Oklahoma newspaper, reported that "'putting the farm on the map' with a permanent number which locates it offhand for the stranger … has been undertaken successfully in Colorado by John B. Plato, an inventive farmer of Bloomfield, near Denver."[30] The article quoted the secretary (not named) of the American Opportunity League as saying, "Right now [Plato] is making a county map in Colorado according to his plan." Although both periodicals misspelled the name of Plato's hometown—it's Broomfield, not Bloomfield—and Internet searching failed to identify the American Opportunity League, Plato was indeed at work on a map, for Fort Collins and vicinity, about 50 miles north of Broomfield.

Ever vigilant about intellectual property, Plato registered a copyright for his "Rural Directory Map, Fort Collins, Colorado." According to the *Catalog of Copyright Entries*, the map was published on 25 November 1916, 16 months after his patent was approved.[31] It was printed on one side of a 36 × 53 cm (14 × 21 inch) sheet of moderately thick paper that was folded into a four-page booklet with the map on the inside and advertising on the title page and back.[32] Immediately below the title, the slogan "'Clock System' Patent" flaunts the patent as a badge of official approval. On the inside, at the lower-left corner of the map, the patent is mentioned again, along with the issue date (7.27.15), just below the notification "Copyrighted 1916 by the U.S.R.D. Co.," identified on the back as the "United States Rural Directory Co., Mountain States Division, Denver, Colorado." The phrases "1916 Edition" and "Revised Yearly" on the front implied consumers would welcome regular updates.

A 33 × 34 cm map shares the sheet with two columns of explanation, one in English and the other in German. The latter's blackletter Gothic type was no doubt familiar to the German-speaking immigrants recruited from Russia early in the century to work in the region's sugar beet fields and factories.[33] Below the English version a simple diagram of sectors and circles illustrates "how the numbers are arranged around a town," and at the center a carefully sketched stem-wound pocket watch reinforces the clock metaphor (Fig. 2.7). Nearby text explains the small black squares

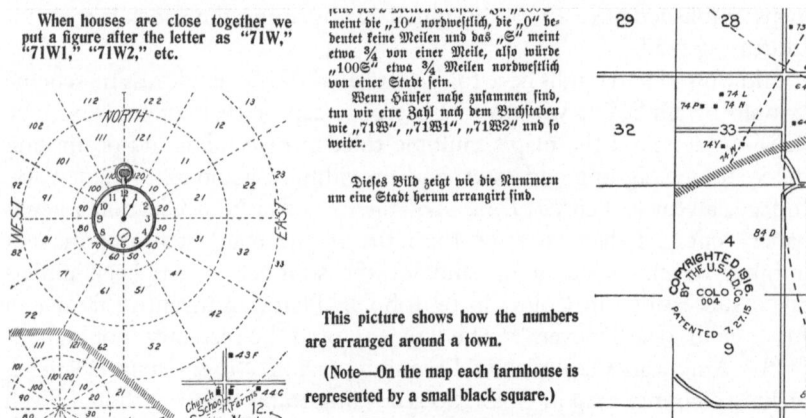

When houses are close together we put a figure after the letter as "71W," "71W1," "71W2," etc.

NORTH

WEST

EAST

This picture shows how the numbers are arranged around a town.

(Note—On the map each farmhouse is represented by a small black square.)

Dividing line between towns   Section Numbers

COPYRIGHTED 1916 BY THE U.S.R.D.C.O. COLO 004 PATENTED 7.27.15

Fig. 2.7 Excerpt from the lower-left part of the centerfold (interior) of the "Rural Directory Map, Fort Collins, Colorado," 1916. Size of original: 53 × 25.4 cm. Size of detail: 23.6 × 12.8 cm. Courtesy Geography and Maps Library, University of Illinois at Urbana–Champaign

that represent individual farmhouses and the thick gray lines that divide the area among Clock System frameworks centered on Fort Collins and three smaller neighbors: Aporte, Harmony, and Timnath.

No bar scale is needed because the circles' one-mile increment and the square-mile sections of the Public Land Survey grid provide an adequate sense of distance. As shown in an excerpt covering the city of Fort Collins and the area directly east (Fig. 2.8), the map's geographic frame of reference includes railways, section-line roads, the Cache La Poudre River, and locally important landmarks such as schools identified by name and the sugar factory. The map's decidedly rural focus is underscored by the omission of residential streets within the one-mile radius and the tagging of each little black square with its Clock System address: one or two digits for the hour, an additional digit for the distance ring, and a unique letter for the structure—a variation covered by Plato's patent. In later directories the areas bounded by two circles and two spokes were called "blocks," like their urban counterparts.

However logical the framework, Clock System addresses were largely useless without the eight-page directory that accompanied the map. The directory's existence is confirmed by the *Catalog of Copyright Entries*, which reported a combined selling price of $0.50 for the map and direc-

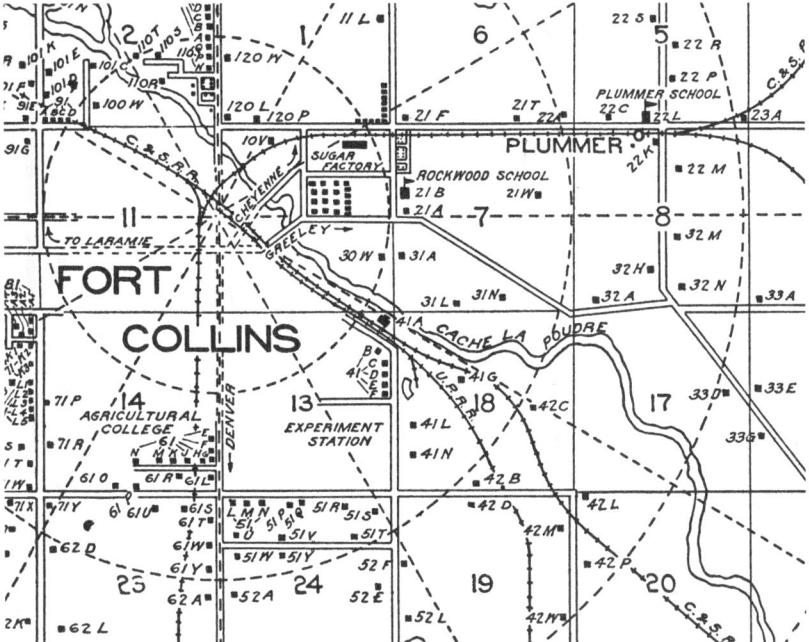

Fig. 2.8 Excerpt from map in the centerfold (interior) of the "Rural Directory Map, Fort Collins, Colorado," 1916. Size of the entire centerfold: 53 × 25.4 cm. Size of detail: 14 × 11 cm. Courtesy Geography and Maps Library, University of Illinois at Urbana–Champaign

tory. The online bibliographic database WorldCat found only one copy of the map, at the University of Illinois in Champaign–Urbana, but its associated directory was missing. The Library of Congress, which runs the Copyright Office, where Plato would have sent two copies, has neither the map nor the directory, and I was unable to find either one online at various Colorado libraries and historical societies. Plato's first published map is apparently rarer than Gerard Mercator's famous 1569 world map, of which three copies are known to exist.[34]

Why Fort Collins, rather than a rural center closer to Denver? Although impressive growth near Fort Collins promised an expanding customer base, keeping the map current would have been troublesome. Perhaps Plato hoped to take advantage of research carried out for the Great Western Sugar Company, which ran the local processing plant and had recruited the area's German-speaking laborers.[35] While searching online for a copy of Plato's directory, I discovered the map "Irrigated Farms of Northern Colorado," commissioned by the company and finished in 1915.[36] Although parcels were labeled with the names of owners, the map did not show farmsteads and other rural residences. Another likely source of data was the US Geological Survey, which had surveyed the area in 1905–06 and issued three editions of its Fort Collins topographic quadrangle map between 1906 and 1909. Its inch-to-a-mile scale (1:62,500) made the map a useful starting point, but it failed to show numerous new farmsteads as well as the Union Pacific Railroad branch that had recently entered town from the southeast.[37]

Compiling a rural directory was no easy task. In addition to checking public records, Plato had to pinpoint every farmstead, which entailed covering countless miles by horse or car. The Fort Collins project also demonstrated the value of enthusiastic support from local public officials, schools, advertisers, and investors, who collectively helped cover start-up costs, provided much of the data, and became the client base. Although Plato's patent described why and how his invention was useful, turning the Clock System into a commercial success demanded further creativity.

Three years elapsed before Plato published his next Clock System directory, for the township of Ulysses, directly northwest of Ithaca in Tompkins County, New York. Unlike its two-part Fort Collins counterpart, the Ulysses directory is a 7½ × 10½ inch, 32-page booklet, in which the map occupies a single page and alphabetical and numerical lists of farmers and their addresses, numerous advertisements, and a detailed discussion of the Clock System and its merits fill the others. Diagrams described how the system worked, and a cartoon captioned "Did you ever try to find 'the third house beyond the little woods on the road next to the old Brown place'?" emphasized why farmers "ought to have house numbers like city folks."[38] Repeated references highlighted the system's usefulness. Ads for local businesses outside a village center typically included the Clock System address, while an image with several typical newspaper want ads demonstrated the comparative advantage of a real address.[39]

No less persuasive were photographs of metal plates with a Clock System address attached to a farmer's mailbox or nailed over the entrance to a rural school.[40] About 16 inches long and 3½ inches tall, these "number plates" were an essential part of the scheme. In addition to making each Clock System address visible on the ground, the plates signified the resident's endorsement. The metal address plates were free, but for $1.50

Fig. 2.9 John Byron Plato. From *Clock System Rural Index, Ulysses Township* (Ithaca, NY: American Rural Index Corporation, 1919), 6

a farmer could order an 18 × 24 inch "bulletin board" like the one pictured in the Ulysses index.[41] Above the number plate a painted panel advertised the names of the farm and its owner. Immediately below a section with "special paint on which to use chalk" accommodated announcements like "fresh eggs today."

That a farmer had invented the Clock System was mentioned repeatedly. The Ulysses index included a photo of Plato standing next to a touring car holding what looks like a map (Fig. 2.9), and the caption highlighted his eagerness to include all farmers.

As explained to readers, the rural directory was a community endeavor focused on farmers, who received not only a free number plate but also a free copy of the rural directory, which sold for $0.50 at banks, drugstores, and garages. By not accepting advertising from mail-order firms outside the area, the directories encouraged residents to shop locally. In an era when it was considered uncouth for professionals to advertise, local physicians, dentists, and attorneys could buy a uniform announcement, laid out like a business card, "without hurting their standing."

Although recruiting school children to collect data for free smacks of exploitation, their involvement was pitched as a unique educational experience. Each rural school was given copies of an enlarged base map of its district to which pupils could add residences and the names of occupants. A facsimile of a school district map attributed to a nine-year-old girl appeared above endorsements from pupils ("Making a map is fun") and their teacher ("A fine lesson in geography").[42] High school students could help the county agent or another local official compile a township map, and once the directory was published, older children could earn a nickel or dime putting up address signs on mailboxes or trees.[43]

In touting the advantages of rural directories while describing how to make one, the rural index for Ulysses was a prospectus promoting wider adoption of the Clock System. The plan included the county agent, who would help select the map's central places, as well as a local printer and a local "bank or club" that would earn the right to advertise on the cover by paying for making the maps and signs.[44] For the Ulysses Index, this honor went to the State Bank of Trumansburg. Much like a fast-food franchise, the "central organization" (Plato's firm) would supply the paper, precut to size, and prepare the press plate for one side of the map, which could be printed on a single sheet and folded. Ideally, the local sponsor would solicit advertising and pay for printing and distribution. In addition to keeping advertising revenue, the sponsor would benefit from the sale of individual copies of the index as well as copies of "enlarged township maps which every store, hotel and garage will want."[45]

Plato's principal backer was the Tompkins County Farm Bureau, listed in copyright filings as co-publisher of the Ulysses directory and five additional Clock System directories covering other rural parts of the county.[46] The first page of each *Rural Index* included an endorsement by the Farm Bureau president, who believed "the Farm Numbering plan will undoubtedly introduce better business methods on the farm" and was confident it "will spread and succeed entirely on its merit and thru local support."[47]

The Ulysses directory included another endorsement on its inside front cover: a facsimile of a letter to Plato from Dr. Liberty Hyde Bailey, retired dean of the agriculture college at Cornell. Bailey had chaired the Commission on Country Life, formed in 1908 by President Theodore Roosevelt to make rural living more attractive.[48] In endorsing the Clock System, Bailey noted that the farmer's home "should be a place on the map and a recognized unit in the community."[49] His letter concluded with the puzzling note, "Of course I have a special interest in your work, remembering that you were my student more than twenty years ago"— puzzling because I could find no connection between Plato, who as far as I know never attended college, and Bailey, whose prior professorial career involved only the Michigan Agricultural College and Cornell.[50]

A "Special Notice" on the inside back cover of the Ulysses directory adds to the mystery of Plato's activities in the years between his Fort Collins and Ulysses maps: "In 1916 and again in 1919 the inventor gave some Chicago men an opportunity to develop the patent but owing to the war their plans could not be carried out, and their connection with the work now has entirely ceased"—I found no records identifying these men, but Chicago was the country's leading center of commercial cartography in the early twentieth century. Equally vague is "the number of conferences [with] representatives" from the Post Office and the Departments of Agriculture and Interior, which led to Plato's decision "that the best and quickest way to get the 'Clock-System' into actual use was to develop a plan whereby the work could be taken up in small units, standardized, and made self supporting"—precisely the plan outlined in the Ulysses directory.

Plato acknowledged the "help and cooperation" of several prominent advisors: Charles A. Lory, president of the Colorado Agricultural College, in Fort Collins; Eugene Cunningham Branson, a rural economist at the University of North Carolina; Albert R. Mann, dean of the agricultural college at Cornell; and Charles Josiah Galpin, a rural sociologist at the US Department of Agriculture. In his 1918 book *Rural Life*, Galpin used a highly generalized Clock System framework centered on Mount Horeb, Wisconsin, to show how "a Colorado farmer in the pure-bred Guernsey business" had "devised an ingenious method for giving every farmer" a business address.[51]

Another supporter was Dwight Sanderson, a professor of "rural social organization" at Cornell, who borrowed several illustrations from the Ulysses directory to highlight the value of not only "putting the farmer on the map" but also giving him an address that reflected his real community. The Post Office had recently closed many rural post offices, often reassigning the farmer to "a rural route starting from some railroad station or larger town which he visits only occasionally."[52] A Clock System address would fix that.

Plato's endeavor caught the attention of *The Literary Digest*, which profiled the Clock System in February 1920.[53] The popular weekly had no need to interview the inventor, whose tale of frustration when composing a newspaper ad for his Guernsey calves was taken verbatim from the Ulysses directory, along with photos comparing rural mailboxes with and without a Clock System address plate and a pair of line drawings that invoked a stem-wound watch to explain addresses centered on a hypo-

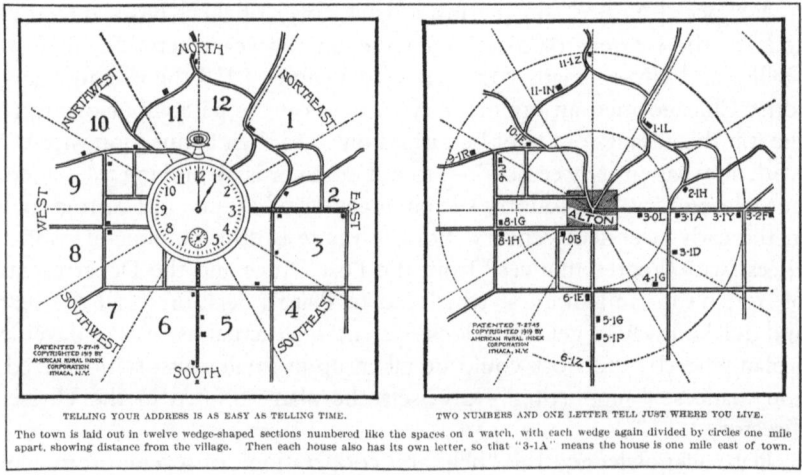

TELLING YOUR ADDRESS IS AS EASY AS TELLING TIME.   TWO NUMBERS AND ONE LETTER TELL JUST WHERE YOU LIVE.

The town is laid out in twelve wedge-shaped sections numbered like the spaces on a watch, with each wedge again divided by circles one mile apart, showing distance from the village.   Then each house also has its own letter, so that "3-1A" means the house is one mile east of town.

Fig. 2.10 Graphic explanation of the Clock System in the 21 February 1920 issue of *The Literary Digest*. Size of original: 18 × 10.1 cm

thetical village (Fig. 2.10).[54] That three-quarters of the short article was in quotation marks confirmed the directory's dual role as a press release.

A year earlier the *Ithaca Daily News* had highlighted the local angle in a story headlined "New Company for Ithaca."[55] No illustrations were included, and the short article said little about how the Clock System worked, but its tone was enthusiastic. The company would "attend to the necessary mapping and publishing and supply suitable signs that are placed in front of every farm house, school, mill or other building in the community," and farmers, who would receive a free copy, could now "advertise in a more businesslike way"—an obvious boon to the local press. "The work of issuing a series of rural directories throughout the eastern states" would benefit the local economy, and with Ithaca's (and Cornell's) reputation for innovation well established, readers could take delight that "the town of Ulysses is the first one in the United States in which the system has been adopted."

The story also reported that the recently chartered American Rural Index Corporation, capitalized at $10,000, would be overseen by Plato and six other directors, all mentioned by name. Incorporation papers indicate that Plato held 15 of the initial hundred shares.[56] Although he was not a Cornell graduate, several of his coterie were, according to a January 1920 article in *Cornell Alumni News*.[57] Charles T. Stagg '02 was company president; Juan E. Reyna '98 was secretary; and Charles E. Treman '89 was treasurer—Reyna, who would have been about Plato's age (43), held half the shares. Non-alumni included Plato, who was vice-president, and executives at the *Ithaca Journal-News*, the First National Bank of Ithaca, and the Ithaca Engraving Company. To build momentum, Plato had become well connected.

His efficient recruitment of backers is impressive. A mid-September 1918 World War I draft registration is the earliest record of Plato in Ithaca, where he was working as a machinist at the Thomas Morse Airplane Company, a key manufacturer of military aircraft.[58] Adept with tools, he had found work to cover living expenses. His mother, Helen, listed as next-of-kin, was still living in Geneva, Illinois, where they had relatives. The 1919 city directory located Plato and his mother at a rooming house in Ithaca, but did not mention his place of employment. Nor did the 1920 Census, for which he reported his occupation as "farmer." Although the 1922 city directory listed him merely as an employee of the "Am Rural Index Journal Corp.," the 1923 directory corrected the name of the firm, listed Plato as its manager, and reported him and his mother at a new address. While in Ithaca, he moved at least two other times.[59]

An expanding cartographic coverage reflects the company's early momentum. In February 1920, Cornell's *Extension News Service* noted that work had begun on the towns of Dryden and Groton, and that "the whole of Tompkins County will be mapped before long."[60] The December issue reported that mapping was underway in Madison County, east of Syracuse. But according to the *Catalog of Copyright Entries*, the map covered only the relatively affluent Town of Cazenovia on the county's western edge. Collaborative compilation by schoolchildren soon faded, along with its intrinsic challenge to quality control. By the end of 1922, the company had copyrighted directories for six counties in New York and one in Pennsylvania, and by the end of 1923, it added two other New York counties (Fig. 2.11). Copyright registration slowed down after the firm changed its name to Index Map Company in 1924. Only three counties were added between 1924 and 1929.

Plato's position as a minority shareholder changed along with the firm's name. According to the new certificate of incorporation, filed in mid-April 1924, he now held 98 of the 100 shares of common stock.[61] His mother, Helen L. Plato, held one share, as did E. Morgan St. John, a local lawyer. Because a board of only three directors governed the new firm, Plato essentially had complete control.

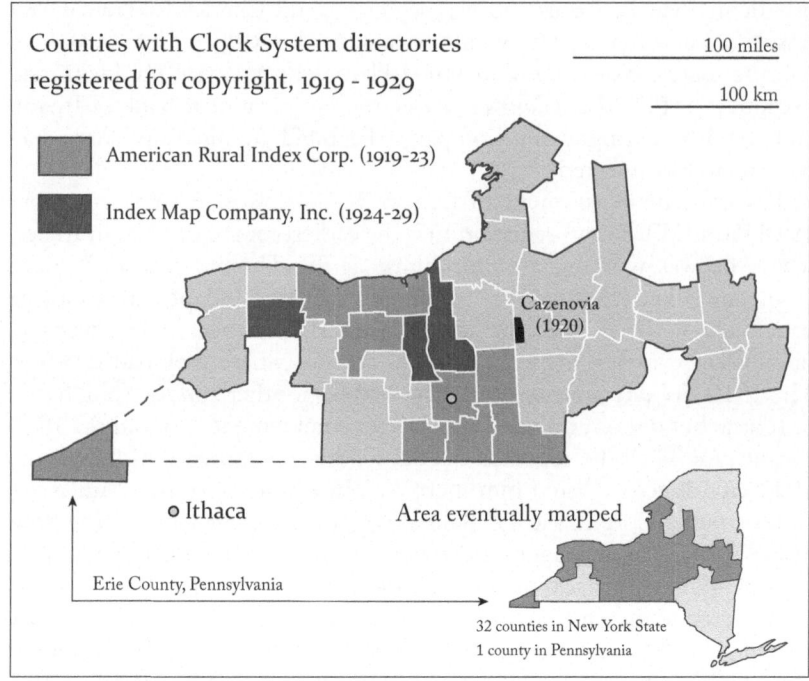

Counties with Clock System directories
registered for copyright, 1919 - 1929

100 miles

100 km

American Rural Index Corp. (1919-23)

Index Map Company, Inc. (1924-29)

Cazenovia
(1920)

○ Ithaca   Area eventually mapped

Erie County, Pennsylvania

32 counties in New York State
1 county in Pennsylvania

Fig. 2.11 Counties with copyrights registered for a Clock System directory, 1919–29. Graytones differentiate coverage by the two companies with which Plato was involved. The lightest graytones show counties covered by a successor firm (see Fig. 2.14), and the inset map shows the entire area eventually covered by a rural directory. Compiled by author from the *Catalog of Copyright Entries*

As the business model evolved, free address plates were dropped and farmers were no longer promised a free copy. A new product emerged around mid-decade, exemplified by the *1926–27 Special Classified and Graded List of All Farms of Broome County, New York, for Merchants and Manufacturers*.[62] Its cover page pitched the *List* less to farmers than to businesses that could use the information for "circularizing," "follow-up work," "deliveries," "billing," and "collections, etc." The alphabetical listing included both postal and Clock System addresses, and letter codes distinguished among general (G), dairy (D), fruit (F), and poultry (P) farms. In addition, the listing graded farms as very large (F), large (A), medium (R), or small (M)—a puzzling code until you realize the letters spell FARM—and a residence that was not a farm was identified with an H, for country home. Across all categories, capital and lower-case letters distinguished owners from tenants, respectively. No price was mentioned for the *List*, which did not include a map, but a "Special County Farm Map" could be purchased for 50 cents. As a reminder that the original client base was still privileged, a "Farm Edition … called a 'RURAL INDEX' … sells to farmers and farm wives at 10 cents."

Variable pricing suggested a willingness to experiment. Two years later and two counties farther north, in Onondaga County, the maps cost 75 cents but the rural index was "distributed free to farmers." By contrast, for Seneca County, just northwest of Tompkins County, the "Map and Index (containing name[s] and Map Number[s] of 2500 Farmers)" sold for $1.00 in 1924, and the "List of Farmers with correct Post Office address *added*" [emphasis mine] was priced at $2.50. Copies preserved in Cornell's rare books department suggested that new editions were not issued every year and that not all revised editions were registered with the Copyright Office.

A company with few employees or little debt can disappear without mention in the local press, as happened to the Index Map Company in 1931 or 1932. The last copyright for a Clock System map—for Genesee County, well west of Ithaca—was registered in mid-July 1929, less than four months before the catastrophic stock market crash that precipitated the Great Depression. The collapse hit agriculture particularly hard, and no doubt undermined sales of advertising and maps pitched to farmers. With his business failing, Plato had little reason to remain in Ithaca after his 89-year-old mother and longtime companion died in early March 1931.[63] *Manning's Ithaca Directory* for 1931 lists John B. Plato, "pres and mgr Index Map Co.," residing at 201 Center Street, along with his widowed mother, but both were missing from the 1932 edition.[64] Oddly, the listing for the Index Map Company survived a year longer.

Despite the country's slow and fretful economic recovery, Ithaca's rural directory business reemerged five years later with a new name, new leadership, and a not-quite-so-new locational framework. Plato, whose patent had expired in 1932, had no apparent association with Rural Directories, Inc., formed in April 1936.[65] Although the corporate directors were all new, the 8-sector Compass System that replaced the 12-sector Clock System was conceptually similar and apparently the only link between the old and new firms—I found no indication that Plato had sold any records, artwork, or furnishings to the new company, or indeed that anything was left to sell.

Fig. 2.12 Excerpt from the folded map accompanying *Rural Index and "Compass System" Map, Oswego County, N.Y.* (Ithaca, NY: Rural Directories Inc., 1938). Size of original: 96 × 61 cm. Size of detail: 11.1 × 9.5 cm. Courtesy Maps and Cartographic Resources, Bird Library, Syracuse University

For people like me, who must pause when translating "at four o'clock" to "a bit south of east," the Compass System is cognitively simpler than its predecessor. A simple compass card provided a convenient logo for covers and title pages, and with four fewer spokes around each center, the maps were less cluttered. Rural addresses were now more straightforward insofar as each road was given a letter, and sequence numbers for farmsteads increased with distance from the central place with no reset at block boundaries. As shown in Fig. 2.12, circles were still a mile apart and gaps in numbering allowed for new dwellings, just like in cities. Plato had adopted a similar numbering scheme for his 1929–30 Clock System map of Genesee County.[66]

A subtler difference between the old and new directories was the roster of company officials to complement the list of advisors. Plato's directories mentioned only one employee—himself, as president—but the 1929 city directory indicated he employed a part-time secretary (who also ran his own realty firm).[67] By contrast, the *Cortland County, N.Y. Rural Index and "Compass System" Map*, published in 1937, listed three people: H. Stillwell Brown (executive vice-president), Carl L. Buchanan (editor), and George R. Hoerner (cartographer).[68] That year's city directory indicated the new firm also employed a bookkeeper and a "field census manager."[69]

A year later the rural directory for Oswego County indicated the official roster had doubled: Brown was now president and general manager, Carl L. Buchanan was vice-president and sales manager, R. M. Engleson was treasurer, Roger S. Reid was field supervisor, the editor was now Joseph A. Short, and the cartographer was now Glen E. Bullock.[70] In late 1938, a corporate restructuring changed the firm's name to Rural Surveys, Inc., and installed Engleson as president, with Brown as vice-president and managing editor.

Compass System directories reflected an increased concern for convenience, content, and graphic design. A map pocket at the back helped keep the folded map with the booklet, and short farm-oriented articles and tables offered advice on issues ranging from controlling insects to estimating the weight of dairy cattle. Layout and artwork had a more polished look, at least by 1930s standards. Several directories explained the addressing system by juxtaposing an enlarged map excerpt with an oblique air photo, annotated with circles and radial lines focused on a local business center (Fig. 2.13). By the late 1930s, most Upstate farmers would have been familiar with the aerial imagery used by the Agricultural Adjustment Administration, a New Deal program concerned with price stabilization and soil conservation.[71] Although block 2E was directly above block 2SE on the folded map, in accord with the traditional north-up orientation, labels in this customized excerpt had been rotated for easy reading.

Updated editions for all but two of the counties mapped in the 1920s accompanied expansion farther east and west (Fig. 2.14). Post-Plato compilation shrewdly avoided the more rugged, less agriculturally prosperous areas in the Catskills (to the southeast) and the northern Appalachians (in the southern tier of western New York counties). No less forbidding was the Adirondacks, to the northeast, where Compass System mapping ignored the more marginally agrarian northern portions of Herkimer and Saratoga counties. In general, Rural Directories, Inc., registered copyrights

Everyone realizes that "John Doe, R.F.D. 3, Farmville" indicates little regarding the exact location of John Doe's residence. To locate a certain individual "who lives on R.F.D. 3" invariably involves considerable guessing and searching. RURAL INDEX & MAP—with its "compass" address of each rural home outside the city and village limits and non-agricultural areas of Oswego County—now eliminates all such uncertainty. *See example as shown in box at bottom of this page.*

To the left: Aerial View taken near Taughannock Falls State Park, Trumansburg. (Camera pointed west.) Radial and circular lines have been superimposed on the photograph to illustrate the "Compass System" of locating rural homes.

Below: Section of Tompkins County "Compass System" Map for part of area shown in photograph above.

Here's the Farmhouse in the Picture–

Here it is on the "Compass System" Map.

This section of a typical county "Compass System" Map shows how the county is divided into zones, each having a city or village as its center. Radiating from this center are eight lines, dividing the surrounding area into zones designated by the four main and four intermediate points of the compass.

Concentric circles, a mile apart, divide the spokes into blocks which are designated as 1NW, 3SE, 2S, etc. ("3SE" indicates a block that is three miles south east of the center from which it radiates.)

the Compass **IS THE KEY**

LEGEND
HIGHWAYS
IMP. ROADS
DIRT ROADS
RAILROADS
HOUSES
COMPASS CENTER

Fig. 2.13 Explanation of the Compass System in the Oswego County directory emphasized the linkage between the map and the physical landscape. *Rural Index and "Compass System" Map, Oswego County, N.Y.* (Ithaca, NY: Rural Directories Inc., 1938), 2. Size of original: 20.9 × 22.1 cm. Courtesy Maps and Cartographic Resources, Bird Library, Syracuse University

for the low-hanging fruit—former Clock System territory (Fig. 2.11) and new counties within a hundred miles of Ithaca—and its corporate successor, Rural Surveys, Inc., registered copyrights for counties farther afield.

Despite (or perhaps because of) its larger staff and more impressive product, the Compass System enjoyed a much shorter run than the Clock

Counties with Compass System directories
registered for copyright, 1936 - 1940

100 miles

100 km

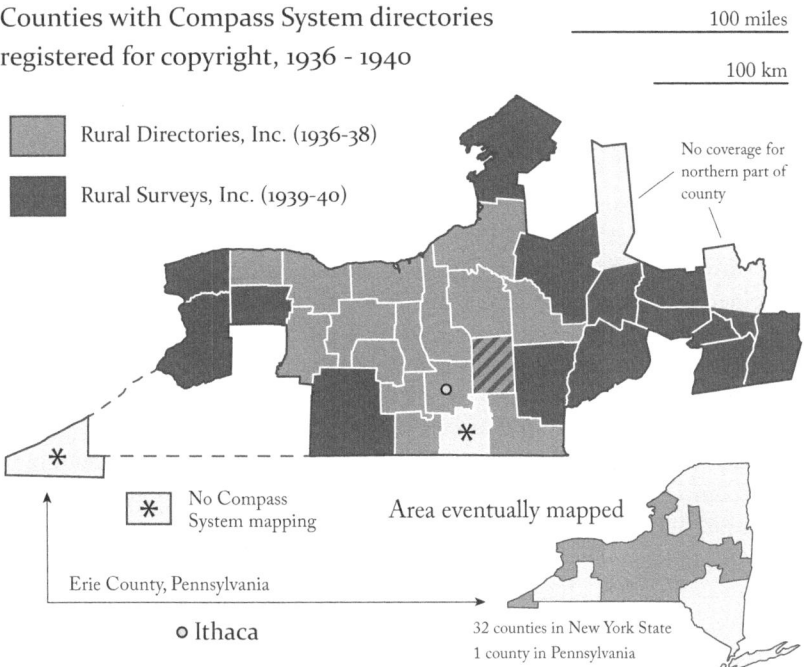

Rural Directories, Inc. (1936-38)

Rural Surveys, Inc. (1939-40)

No coverage for
northern part of
county

* No Compass
System mapping

Area eventually mapped

Erie County, Pennsylvania

o Ithaca

32 counties in New York State
1 county in Pennsylvania

Fig. 2.14 Counties with copyrights registered for a Compass System directory, 1936–40. The inset map shows the entire area eventually covered by a rural directory. Of the counties mapped between 1919 and 1929 (Fig. 2.11), only Tioga County, south of Ithaca, and Erie County, Pennsylvania, were not covered by a Compass System directory. Compiled by author from the *Catalog of Copyright Entries*

System, which had lasted more than a decade. In 1940 or 1941, Rural Directories/Surveys expired after what human obituaries call a brief illness. In 1940 an ostensibly healthy Rural Surveys, Inc. registered copyrights for directories covering five counties, the last in mid-May, for Otsego County, about 60 miles east of Ithaca.[72] Even so, signs of a failing business are apparent in the 1940 Ithaca city directory, which reported the company presidency "vacant" and listed H. Stillwell Brown as filling three roles: executive vice-president, editor, and treasurer.[73] By 1941 the company had moved from a prominent downtown location (147–151 East State Street) to Brown's home, at 945 Cliff Street, and Brown himself was now a salesman for a local radio station.[74] And by 1941 Rural Surveys, Inc. was gone completely, the likely victim of overly optimistic expansion, with revenues never catching up with expenses—an endemic ailment unlike the economic pandemic that undermined Plato's more conservative endeavor a decade earlier. As a business plan, the Compass System had failed the market test.

Like other casualties of the Great Depression, Plato found refuge with the federal government, in Washington, DC, where the 1935 city directory shows two different but equally plausible listings: one for J.B. Plato, identified as a "spl agt" [special agent?] with the Census Bureau living in Forestville, Maryland, and another for "John B Plato," an assistant agricultural economist with the Agricultural Adjustment Administration living in the District's Benning neighborhood, in roughly the same direction as Forestville but closer to downtown.[75] The 1936 edition confirmed that he was a Forestville resident, but a 1937 Associated Press story placed him in rural Louisiana, helping the Department of Agriculture coordinate a cartographic directory for an area recently ravaged by a tornado.[76] Without mentioning the Compass System by name, the AP story described a cartographic framework identical to Rural Directory mapping then underway in New York State.

When Plato died in 1966, at age 89, his *Washington Post* obituary noted "he came to Washington in 1933 as a map expert for the government," but said nothing about his patent, rural directories, or experiences in Ithaca.[77] Instead, the obit emphasized the nature preserve he had set up in Forestville, his marriage in 1948 (to a woman 24 years younger), and his donation of 50 acres of woodland for a Girl Scout camp—further examples of the initiative and resilience behind the Clock System.

As Plato's story demonstrates, the success of a cartographic invention depends on far more than a clever idea and a government patent. Not all patentable inventions are marketable, not all inventors have the ingenuity needed to recognize and develop a market, and not all potential markets are stable and easily maintained. Success—survival might be a better word—depends upon a flexible business plan, the emotional commitment of the inventor and his backers, and the health of both the national economy and locally or regionally important sectors, such as agriculture. And as Plato's experience demonstrates, the possibilities for success can be limited in both space and time. Although the Clock System might have been equally useful in, say, rural Wisconsin, map-based rural directories never went national, probably because their innate utility was too narrow to take root without the enthusiastic presence of John Bryon Plato, the rural sociologists at Cornell, and a handful of Ithaca businessmen intrigued by a clever cartographic solution to the momentary difficulties of inadequate rural telephone service and poorly paved roads with incomplete signage.[78]

NOTES

1. Samuel E. Gross, "Map," US Patent 232,261, filed 16 May 1878, and issued 14 September 1880.

2. If not already available, the prerequisite technology was very much in the wind. See, for example, T. Seymour Scott, "Improvement in the Manufacture of Flexible Paper," US Patent 216108, filed 8 February 1879, and issued 3 June 1879.

3. William A. Baugh, "Method of Subdividing and Designating Land," US Patent 367,178, filed 3 January 1887, and issued 26 July 1887.

4. For a history of the Public Land Survey System and the numbering of sections, see Hildegard Binder Johnson, *Order upon the Land: the U.S. Rectangular Land Survey and the Upper Mississippi Country* (New York: Oxford University Press, 1976), esp. 53–64; and Gaby M. Neunzert, *Subdividing the Land: Metes and Bounds and Rectangular Survey Systems* (Boca Raton, FL: Taylor & Francis, 2011), esp. 21–27. For a history of quadtree data structures, see Donna J. Peuquet, "Electronic Cartography: Data Structures and the Storage and Retrieval of Spatial Data," in *Cartography in the Twentieth Century* (Vol. 6 of *The History of Cartography*), ed. Mark Monmonier, 359–66 (Chicago: University of Chicago Press, 2015). Peuquet attributes early interest in quadtrees to a book chapter by Allen Klinger, "Patterns and Search Statistics," in *Optimizing Methods in Statistics*, ed. Jagdish S. Rustagi, 303–37 (New York: Academic Press, 1971). Although Baugh's invention had the successive hierarchical subdivision characteristic of the quadtree, it lacked the numerical topological "tree" structure used to describe the nested hierarchy.

5. John A. Jones, "Geographic Location Identification System," US Patent 5,445,524, filed 3 May 1994, and issued 29 August 1995.

6. "US Patent No: 5,445,524: Geographic Location Identification System," PatentBuddy, http://www.patentbuddy.com/Patent/5445524.

7. Jeff A. Ronspies, "Comment: Does David Need a New Sling? Small Entities Face a Costly Barrier to Patent Protection," *The John Marshall Law School Review of Intellectual Property Law* 4 (2004): 184–211, esp. 194.

8. For a concise overview of the adverse consequences of excessive premature patenting, see Daniel Porter, "Dormant Patents," Patexia Community blog, 30 January 2013, https://www.patexia.com/feed/dormant-patents-20130204#.

9. Moderate online searching as well as a query to the US Patent and Trademark Office failed to uncover the exact maintenance fee assessed in 2003 for patents already in force 7½ years. David Fitzpatrick, of the Office of Patent Planning and Capacity Analysis (email, 12 March 2015), provided a link to historical data compiled by Five IP Offices, a consortium of intellectual property offices in Europe, Japan, Korea, the People's Republic of China, and the United States; see http://www.fiveipoffices.org/statistics/statisticaldata.html. A table for the United States provides yearly information for the period 2005 through 2014. The 7½ year fee was $2,300 for 2005–07, $2,380 for 2008, $2,480 for 2009–11, $2,850 for 2012, and $3,600 for 2013–14. In the absence of better data, "more than $2,000" seems a safe estimate. A footnote in the table indicated that on 1 July 2005, the fee was reduced by 50 percent for "small entities"—too late for Jones, whose patent had expired because he did not pay the maintenance fee.

10. John Byron Plato, "Map or Chart," US Patent 1,147,749, filed 7 December 1914, and issued 27 July 1915. On the same date, *The Official Gazette of the U.S. Patent Office* listed the first five of the patent's six claims and reported its assignment to "Cl. 35—6," identified as "Educational Appliances—Maps" in *Manual of Classification of Subjects of Invention of the United States Patent Office, Revised to January 1, 1916* (Washington, DC: Government Printing Office, 1916), 38.

11. Correspondence related to individual patents is organized by patent number as part of Record Group 341.3, stored at the National Archives site in Kansas City, Missouri. John B. Plato to Commissioner of Patents, 3 December 1914.

12. L. D. Underwood to John Byron Plato, 16 January 1915.

13. Ancestry Library Edition led to various city directory references and related genealogical sources, including the 1900 Census, for

which Rollandet reported his occupation as draughtsman as well as his immigration in 1888 and marriage in 1895, and Findagrave. com, which says he died in 1938. The half-page ad was on page 323 of the directory published by Bellenger & Richards, which produced directories for nearly a hundred US cities.

14. *Representative Men of Colorado in the Nineteenth Century: A Portrait Gallery of Many of the Men Who Have Been Instrumental in the Upbuilding of Colorado, Including Not Only the Pioneers, but Others Who, Coming Later, Have Added Their Quota, Until the Once Territory is Now the Splendid State* (Denver: Rowell Art Publishing Co., 1902), 133.

15. Rollandet to Commissioner of Patents, 25 February 1915.

16. Underwood to Rollandet, 31 March 1915.

17. John H. Robinson, "Tariff Chart," US Patent 1,029,085, filed 29 April 1912, and issued 11 June 1912. The other two patents mentioned in Underwood's letter are Allen B. Maull, "Chart or Map," US Patent 1,110,217, filed 21 May 1912, and issued 8 September 1914; and Joel Rindall (assignor of one-half to Henry N. Segerstrom), "Guide-Chart," US Patent 1,039,322, filed 21 October 1911, and issued 24 September 1912.

18. Rollandet to Commissioner of Patents, 8 May 1915.

19. Underwood to Rollandet, 25 May 1915.

20. Rollandet to Commissioner of Patents, 3 June 1915.

21. Commissioner of Patents to Plato (and Rollandet), 16 June 1915.

22. Rollandet to Commissioner of Patents, 22 June 1915.

23. In order of filing, the three patents were "Horse-Hitching Device," US Patent 823,964, filed 21 June 1905, and issued 19 June 1906; "Hitching Device," US Patent 986,591, filed 26 December 1905, and issued 14 March 1911; and "Hitching Attachment for Vehicles," US Patent 807,047, filed 29 April 1905, and issued 12 December 1905. Rollandet represented Plato for the last two applications. The third patent listed a co-inventor, William H. Kilgore, but the award noted that Kilgore had assigned his rights to Plato.

24. The most relevant detailed maps I could find were scanned for the ProQuest database Digital Sanborn Maps 1867–1970. The 1903–04 Sanborn fire-insurance atlas for Denver pinpoints the location of the factory, 1216 Arapahoe, in a mixed-use residential-industrial neighborhood next to a paint factory and across the street from

a cabinet maker. Plato's residence at 2416 Williams St., Denver, is in a neighborhood not covered by Sanborn, possibly because it was quite new at the time. Distance was determined with Google Maps.

25. Ancestry Library Edition was a valuable portal to city directories, Census registers, and other documents too numerous to mention individually. Sources not accessed through Ancestry include his obituary "Uncle John Plato, Benefactor of Girl Scouts in Forestville," *Washington Post*, 1 July 1966, B8.

26. Arthur C. Johnson, *Official History of the Operations of the First Colorado Infantry, U. S. V. in the Campaign in the Philippine Islands* (no publisher identified, 1899), 57.

27. Clyde A. Hall and John B. Plato, "Reducing Barn Costs—Plans Used by Two Dairymen," *System on the Farm* 1.5 (July 1917): 192–93. According to Worldcat.org, the A. W. Shaw Company, in Chicago, published the monthly agricultural magazine from 1917 to 1921.

28. "Transfers of Guernsey Cattle From March 1 to March 14, 1918," *Guernsey Breeders' Journal* 13 (1918). Semi-monthly transfers lists, not paginated, appear toward the end of the volume. In the scanned copy found on the HathiTrust website, the transfer of Flossie is on page 498 of the 528-page PDF document.

29. W. F. French, "Watch Locates Neighboring Farmers," *Illustrated World* 27 (1917): 247–48. The article incorrectly reports his post office as Bloomfield, not Broomfield. Colorado apparently had no Bloomfield. Moreover, his direct correspondence with the Patent Office shows his address as Featherstone Farm, Route 1, Box 112, Broomfield, Colorado, but the application describes him as a resident of nearby Semper, Colorado. As a further complication, the 1916 Denver city directory places him at 1730 Speer Blvd., less than a block from the former Plato Manufacturing Company.

30. "Name or Number the Farm," *Daily Ardmoreite* [Ardmore, Okla.], 26 September 1916, 8; in NewspaperArchive.com.

31. Library of Congress, Copyright Office, *Catalog of Copyright Entries*, Part 1: Books, Group 2, n.s., vol. 14, no. 1 (Washington, DC: Government Printing Office, 1917), 318.

32. United States Rural Directory Co., Mountain States Division, "Rural Directory Map, Fort Collins, Colorado," Denver, 1916.

33. Fort Collins Museum of Discovery, "Germans from Russia," Fort Collins History Connection, 2012, http://history.fcgov.com/archive/ethnic/german.php.

34. Joaquim Alves Gaspar and Henrique Leitão, "Squaring the Circle: How Mercator Constructed His Projection in 1569," *Imago Mundi* 66 (2014): 1–24.

35. Leonard J. Arrington, *Beet Sugar in the West: A History of the Utah-Idaho Sugar Company, 1891–1966* (Seattle: University of Washington Press, 1966).

36. Copies of the "Map of Irrigated Farms of Northern Colorado, 1915," compiled by R. W. Gelder, are in the collections of the Greeley Museums Municipal Archives and the Local History Archive of the Fort Collins Public Library. See "Historical Maps," Fort Collins History Collection, http://history.fcgov.com/.

37. US Geological Survey, Fort Collins, Colorado [quadrangle map], 1:62,500, 15-minute series, editions published 1906, 1908, and 1909.

38. *Clock System Rural Index, Ulysses Township* (Ithaca, NY: American Rural Index Corporation, 1919) [hereafter *Ulysses Rural Index*], 2.

39. *Ulysses Rural Index*, 10.

40. *Ulysses Rural Index*, 4, 8.

41. *Ulysses Rural Index*, 10.

42. *Ulysses Rural Index*, 8.

43. *Ulysses Rural Index*, 10.

44. *Ulysses Rural Index*, 6.

45. *Ulysses Rural Index*, 10.

46. According to the *Catalog of Copyright Entries*, in 1919, John B. Plato filed a copyright registration in his name for a separate Clock System map of Ulysses (published on 20 January 1919), and the American Rural Index Corp. later registered a copyright for the rural index for Ulysses (published 10 December 1919). In 1920 copyrights registered for five rural indexes listed the Tompkins County Farm Bureau as the author and the American Rural Index Corp. as the copyright holder; these indexes covered Danby and Newfield Towns (published 1 November 1920), Dryden Township (18 October 1920), Groton Township (published 1 October 1920), Ithaca and Enfield Towns (published 25 October 1920), and Lansing Township (1 November 1920).

47. George R. Fitts quoted in the *Ulysses Rural Index*, 1.

48. See Clayton S. Ellsworth, "Theodore Roosevelt's Country Life Commission," *Agricultural History* 34 (1960): 155–72.

49. *Ulysses Rural Index*, inside front cover. Bailey's letter is dated 3 November 1919.

50. For examples, see Harlan B. Banks, *Liberty Hyde Bailey, 1858–1954: A Biographical Memoir* (Washington, DC: National Academy of Sciences, 1994); and "Liberty Hyde Bailey Timeline" [interactive online timeline], http://www.timetoast.com/timelines/liberty-hyde-bailey; and Cornell University, Division of Rare & Manuscript Collections, "Liberty Hyde Bailey: A Man for All Seasons" [interactive online timeline], http://rmc.library.cornell.edu/bailey/timeline.html.

51. Charles Josiah Galpin, *Rural Life* (New York: Century, 1918), 341–42; map on 344. The title page associated Galpin with the agricultural economics faculty at the University of Wisconsin.

52. Dwight Sanderson, "Locating the Rural Community," *Cornell Reading Course for the Farm*, lesson 158 (June 1920): 415–36, esp. 429–33; quotation on 433. Sanderson also praised the Clock System in his *The Farmer and His Community* (New York: Harcourt, Brace and Co., 1922), 231–32.

53. "Putting the Farmer on the Map," *Literary Digest* 64.8 (21 February 1920): 28–29.

54. *Ulysses Rural Index*, 4, 12, 14.

55. "New Company for Ithaca," *Ithaca Daily News*, 28 February 1919, 3.

56. Certificate of Incorporation of American Rural Index Corporation, recorded 20 February 1919, Tomkins County, New York; County Clerk control number BF037300-001.

57. "The Rural Index," *Cornell Alumni News* 22.17 (22 January 1920): 194–95.

58. John Byron Plato, World War I Draft Registration Card, 12 September 1918. Plato's draft registration and city directory listings were accessed through Ancestry Library Edition. The correct name of his employer was the Thomas-Morse Aircraft Corporation; see Fay L. Faurote, ed., *Aircraft Yearbook, 1919* (New York: Manufacturers Aircraft Association, 1919), 240–69.

59. According to city directories and other sources, Plato lived at 319 East Mill Street (now East Court Street; 1918 through 1922), 201 Center Street (1923), 514 S. Cayuga Street (1925), and 403 S. Cayuga (1927 through 1931). All were rooming houses or multiple-occupancy residences, and all were within easy walking distance to the firm's office on East State Street.

60. "Tompkins County Farms Are Being Numbered," *Extension News Service* [New York State College of Agriculture at Cornell University] 3.2 (February 1920): 23.

61. Certificate of Incorporation of Index Map Company, Inc., recorded 19 April 1924, Tomkins County, New York; County Clerk control number BF037450-001. A total of 2,000 shares were authorized, half in common stock and half in preferred. Common shareholders in the old American Rural Index Corporation, might have become preferred shareholders, who had no voting rights but earned an annual dividend of 8 percent on the par value of $10 per share. The local historical society has no record of either company, and after 1924 Plato might have been the firm's only employee.

62. *1926–27 Special Classified and Graded List of All Farms of Broome County, New York, for Merchants and Manufacturers* (Ithaca, NY: Index Map Co., 1926), 1.

63. "Mrs. Helen Larrabee Plato" [obituary], *Ithaca Journal-News*, 5 March 1931, 5. Plato's mother had died the preceding day at Memorial Hospital and was to be buried "in the family plot in Geneva, Illinois." Plato was her only child.

64. *Manning's Ithaca (New York) Directory, for the Year Beginning January 1931* (Schenectady, NY: H. A. Manning, 1931), 381.

65. An incorporation date of 18 April 1936 is noted in Certificate of Change of Name of Rural Directories, Inc. to Brown—Engleson Publishing Corporation, recorded 17 October 1938; County Clerk control number BF104963-001. The new name apparently reflects the firm's two principal investors.

66. *"Clock System" Rural Index and Buying Guide for Genesee County for 1929 and 1930* (Ithaca, NY: Index Map Co. 1929). Also see the excerpt from the soon-to-be-published Oneida County map that illustrated the news article "Numbering Oneida County Rural Homes by the Clock System Will Aid in Locating Farms," *Rome Daily Sentinel*, 28 August 1928, 9. Also see the "Improved 'Clock System' Map of Oneida County" (Ithaca, NY: Index Map Co., 1928).

67. *Manning's Ithaca (New York) Directory, for the Year Beginning January 1929* (Schenectady, NY: H. A. Manning, 1929), 387. Frank L. Tyler was listed as "sec Index Map Co and prop The Lion Realty Co." Tyler's listing in the 1932 directory included only his own firm.

68. *Cortland County, N.Y. Rural Index and "Compass System" Map* (Ithaca, NY: Rural Directories Inc., 1937), 3.

69. *Manning's Ithaca (New York) Directory, for the Year Beginning January 1937* (Schenectady, NY: H. A. Manning, 1937), 152, 212. Helen Crispell was the bookkeeper, and Thomas Kirk was the field census manager.

70. *Rural Index and "Compass System" Map, Oswego County, N.Y.* (Ithaca, NY: Rural Directories Inc., 1938), 1.

71. Mark Monmonier, "Aerial Photography at the Agricultural Adjustment Administration: Acreage Controls, Conservation Benefits, and Overhead Surveillance in the 1930s," *Photogrammetric Engineering and Remote Sensing* 76 (2002): 1257–61.

72. Library of Congress, Copyright Office, *Catalog of Copyright Entries*, Part 1: Books, Group 2, n.s., vol. 37, no. 6 (Washington, DC: Government Printing Office, 1940), 671.

73. *Manning's Ithaca (New York) Directory, for the Year Beginning January 1940* (Schenectady, NY: H. A. Manning, 1940), 265.

74. *Manning's Ithaca (New York) Directory, for the Year Beginning January 1941* (Schenectady, NY: H. A. Manning, 1941), 124, 272.

75. Although the online 1935 city directory for Washington, DC, provided by Ancestry Library Edition (Ancestry.com) does not include a title page naming the publisher, advertising suggests the directory was published by the R. L Polk Company; listings for Plato are on 1569. Earlier editions did not list Plato.

76. "Numbers Used for Directory of Farm Area," *Rochester Democrat and Chronicle*, 29 February 1937, 15A; and "New Directory for Farms To Be Developed," *Utica Observer*, 29 February 1937. Scanned clippings of these articles were found on FultonHistory.com, which focuses on historical newspapers in Upstate New York.

77. "Uncle John Plato, Benefactor of Girl Scouts in Forestville," *Washington Post*, 1 July 1966, B8.

78. The percentage of US farms with telephones rose to 39 percent in 1920 but declined thereafter to 25 percent in 1940. See Claude S. Fischer, *America Calling: A Social History of the Telephone to 1940* (Berkeley: University of California Press, 1992), 92–106, esp. 93.

CHAPTER 3

Showing the Way

New technologies often inspire related inventions: among the better known examples are the telephone, which led to the answering machine and ultimately the cell phone, which in turn triggered an explosion of apps. Few inventions transformed the twentieth century as radically as the automobile, which demanded better roads, accelerated the expansion of cities, and changed the ways people shop, socialize, and recreate. By expanding the motorist's geographic range, the automobile created a niche for more convenient cartographic products, most notably the strip map, which molded a route map to fit an elongated sheet of paper. Because early automobiles moved sufficiently rapidly to make any distraction a hazard, a few clever entrepreneurs simplified map reading by moving the strip through a viewing window as the vehicle advanced. By revealing only the most immediately relevant part of the route, they anticipated the GPS navigator by eight decades.

Strip maps were not new. When challenged to identify a prototype, map historians typically point to a fourth century Roman map that describes a road system extending from Britain to India. The original survived only as a twelfth- or early-thirteenth-century copy of an elongated map of the road and its corridor, drawn on parchment in sections glued together to form a 675 × 34 cm scroll named the Peutinger map after the German scholar Konrad Peutinger (1465–1547), who inherited it in 1508.[1] A later exemplar of this long, narrow format is the "Ribbon Map Of The Father

© Mark Monmonier 2017
Mark Monmonier, *Patents and Cartographic Inventions*,
Palgrave Studies in the History of Science and Technology,
DOI 10.1007/978-3-319-51040-8_3

Of Waters," a 336 × 7 cm hand-colored map of the Mississippi River published in St. Louis in 1866.[2]

Publishers of guidebooks and travel narratives readily recognized the advantages of a long, thin map or simply a geographically ordered list of places or waypoints for a particular route. In the late nineteenth century, the League of American Wheelmen, a national membership organization for cyclists, began publishing road books with tables that divided routes into sections, each described verbally (rather than graphically) by its length, gradient, surface material, and condition.[3] Whether these narratives can be called strip maps is debatable, but spatial sequencing makes them at least marginally cartographic. By contrast, *Outing* magazine, aimed at cyclists and other sportsmen, enriched every issue with cycling tours illustrated with conventional, graphic strip maps that were more generalized and less detailed than the textual maps in the Wheelmen road books.[4] When the automobile began to eclipse the bicycle in the early twentieth century, motorist organizations started publishing sets of strip maps or route cards: a do-it-yourself approach to a customized road atlas.[5] In 1937 the American Automobile Association introduced the Triptik, a bound sequence of small, standardized road maps assembled by a travel planner for a specific journey and annotated using rubber stamps to highlight construction zones, detours, and speed traps.[6]

By the end of the nineteenth century, inventors interested in transportation mapping recognized the value of the Peutinger format. Although loose scrolls could be cumbersome, a pair of rollers in a rigid frame made it easy to follow a route laid out on a strip map. If east was at the top of a map for an east-to-west route, the scroll could be wound upward when traveling west or downward when traveling east. And when a serpentine route was mapped onto a narrow strip, cardinal direction was less important than whether the traveler was headed from place A to place B or from B to A. Akin to the railroad maps that had straightened out winding routes several decades earlier, strips maps for motorists and cyclists sacrificed geographic shape for convenience.[7]

Helping cyclists navigate a route was the obvious goal of Harry C. McCafferty, a Montclair, New Jersey, bookkeeper who patented a "Map Holder" in 1898.[8] He illustrated his invention with three straightforward drawings (Fig. 3.1): a perspective view showing a device that could be clamped to a bicycle handlebar, a side view showing a scroll-like strip map mounted on two parallel rollers, and a view from above showing the knobs used to advance the scroll. Between the rollers a small, oblong

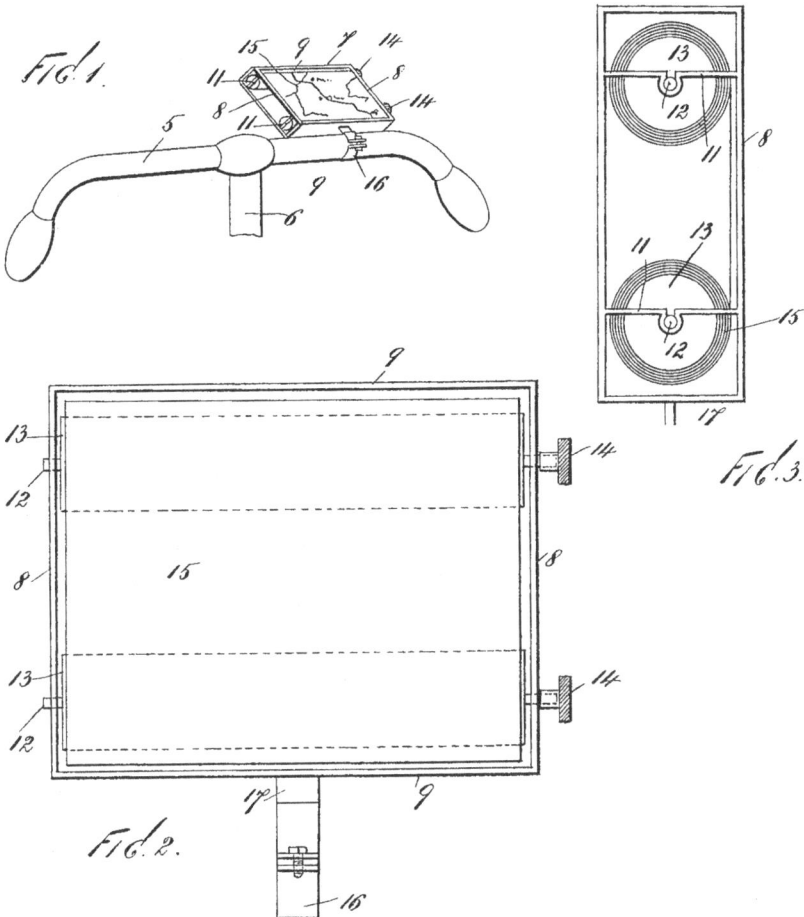

Fig. 3.1 Harry McCafferty used three drawings (rearranged slightly here) to describe his "Map Holder" (US Patent 605,969; 1898)

portion of the map is visible in a rectangular opening, presumably open to the elements. With the scroll (item 15 at the upper right) described as "an ordinary road map," his invention was strictly a "map-holding device," intended for "bicycles or similar vehicles."

Five months behind McCafferty, New York City residents John Kelso and Peter Wilbur filed paperwork for a comparable invention, titled

"Holder for Road-Maps." Like McCafferty's device, it was attached to the handlebar and intended for "bicycles and other vehicles" (but presumably not motor cars).[9] The key difference is a convex viewing window protected from rain by a "transparent plate of glass, mica, or other suitable material." This improvement apparently demonstrated sufficient originality for a patent, awarded 19 months after filing, in May 1899.

Kelso and Wilbur were an odd collaboration. According to that year's city directory and coding sheets for the 1900 Census, the 33-year-old Wilbur was a machinist, and the 62-year-old Kelso worked in real estate.[10] Kelso's contribution probably included an interest in technology and prior experience with the Patent Office: in 1883 he had patented an "Incandescing Electric Lamp"—an unsung rival to the light bulb Thomas Edison had patented three years earlier.[11] They might have tried to sell or license their map-holder patent to someone better able to manufacture the device and supply the cartographic content. A 1911 article in *Scientific American* described a "route-guide … fitted to [an automobile's] steering column and operated by hand" using a single large knob; in an accompanying photograph, it looks similar to Kelso and Wilbur's invention, but the magazine did not name the product, the inventor, or the manufacturer.[12] Inventions patented far outnumber those developed commercially.

Responding to the now-obvious need for ribbon-like maps, Lincoln J. Carter, of Chicago, patented a map useful to cyclists, motorists, and walkers eager "to dispense with the usual heavy cumbersome section-maps at present in use."[13] His patent, awarded in September 1904 and titled merely "Road Map," described an invention consisting "principally … of a strip of flexible material" with a "main heavy line" down the center, intersected by evenly spaced "transverse lines" to show distance as well as landmarks and "branches indicating the turn of the road," as in Fig. 3.2. In his preferred implementation, transverse lines a mile apart were useful for estimating visually the distance between waypoints. But because the route on the ground typically consisted of straight stretches of decently paved road joined at right angles—a consequence of the rectangular land survey, particularly prominent in the Midwest—he added combinations of letters and numbers like "S 8 M." and "E 2½ M." to indicate a route that ran 8 miles due south before heading due east for 2½ miles. In addition, labeled point symbols identify prominent landmarks like the lumberyard and poor farm encountered shortly after leaving Chicago for St. Louis. Dots alongside the road representing "telegraph-poles or trees" provide further guidance, as do wiggly lines for sinuous portions of the route.

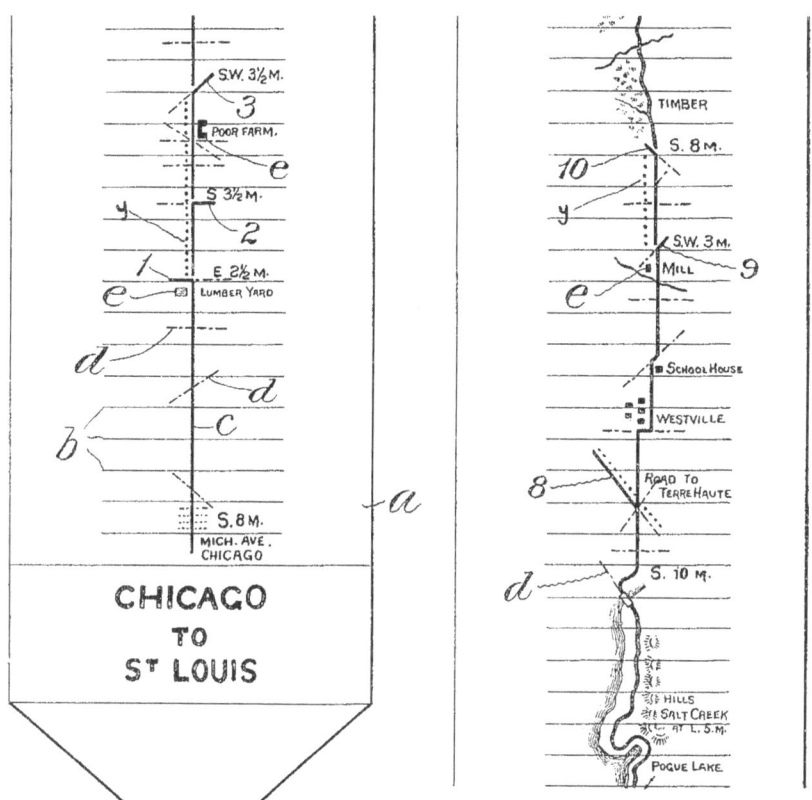

Fig. 3.2 Lower part of the two drawings, published side by side on the first page of Lincoln Carver's "Road Map" (US Patent 770,350; 1904)

Carter avoided any mention of a map holder, aside from the motorist's hand. "It will be seen," he wrote in stilted patentese, "that the principal advantages … of a map constructed in accordance with these improvements is that it can be made of a very small size and carried in a vest-pocket, that only a small portion of it need be exposed at a time, and that the tourist with such a map can travel a road in perfect safety night or day and need not ask any questions regarding the same." That he did not mention a map holder seems odd. St. Louis is about 300 miles from Chicago along today's Interstate 55, and no doubt farther back then. If the map scale was a stingy six miles to the inch, the resulting five-foot-long strip map could interfere with the driver's legs or flap around in the wind. Perhaps he was leery of infringing the claims of inventors like McCafferty, who had recognized the advantage of a pair of spools.

Aside from several other patents, none related to road maps, Lincoln Carter seems an unlikely cartographic inventor. He was born in Rochester, New York, on 15 April 1865, the day after President Abraham Lincoln was assassinated. For the 1900 Census, he reported his occupation as theatrical manager, and that year's Chicago city directory listed him as "prop[rietor] Court theatre." Patents titled "Theatrical Appliance" (1900), "Stage Appliance for Theaters" (1901), "Theatrical Scenery" (1902; 1903; 1907), and "Fireproof Curtain" (1906) reflect experience with the Patent Office as well as a creative impulse.[14] According to the website MoviesPictures. org, Carter's "realistic staging of such dramatic events as train and ship wrecks were only surpassed by the advent of motion pictures."[15] He surely had an appreciation of graphics and a well-developed sense of spatiality. Another plausible influence was Chicago's vigorous map trade.[16]

The obvious next step was a device that not only integrated the ribbon-like route map with the spool-based map holder but also advanced the map automatically. Indeed, tying a scroll-like display to the motion of a vehicle was hardly a new idea in the early twentieth century: in 1880 William F. Johnson, of Providence, Rhode Island, had patented a mechanical contraption that harnessed a roll of appropriately sequenced station names to the movement of a streetcar or subway train. As the drawings for his "Automatic Station Indicator and Advertiser" (Fig. 3.3) show, the rotation of a flanged wheel is transmitted to the rollers and scroll by a connecting rod, a ratchet wheel, a vertical shaft, and bevel gears on the shaft and rollers.[17] As the car moves forward, the scroll rolls upward, just like clockwork. But close inspection reveals that Johnson's "band or sheet [with] the names of streets or stations ... and also advertisements" is not a conventional, double-ended scroll, but a short continuous belt—too short to accommodate most strip maps.

Although nothing in the patent suggests an automated road map—hardly surprising because the motorcar was yet to come—this inconsistency did not stop a patent examiner from citing Johnson's cleverness when questioning the originality of an invention conceived a quarter century later by Frank J. Lindenthaler, of Glens Falls, New York, and John Protz, a citizen of Austria-Hungary living in New York City.[18] According to coding sheets for the federal census, both were naturalized citizens born in Austria and working as machinists, both had a German-speaking

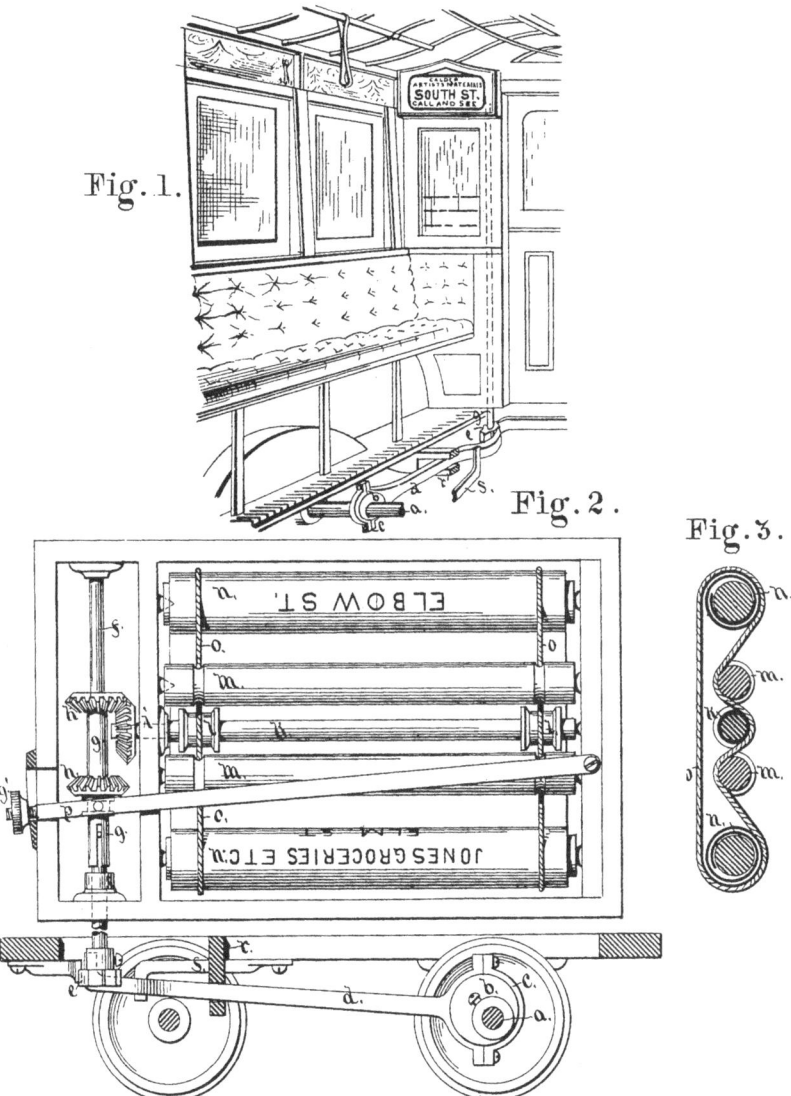

Fig. 3.3 William Johnson's apparatus linked a scroll with station names to the movement of a railway car (US Patent 231,961; 1880)

wife and several children, and both seem to have assimilated quickly to their new country.[19] Lindenthaler, who was born in 1878, had arrived in the United States in 1888, and lived at various times in New York City, in Hudson, New Jersey (across the river), and in the vicinity of Glens Falls, New York, about 200 miles north. In registering for the World War I draft in 1918, he reported living in the Bronx and working as an auto mechanic. By contrast, Protz was 20 years older and a more recent immigrant: he was born in 1858 and arrived in the United States in 1902, apparently from Romania, where his six children had been born. Like Lindenthaler, he worked as a machinist. Both apparently had sufficient technical knowledge to collaborate on the "Road Indicator for Automobiles," for which they filed a patent application in November 1907. I have no idea how they met, but this was the only US patent for either of them.

Their patent describes a strip map mounted on two reels within a metal casing and a set of gears, belts, cams, pawls, and pulleys that advance the map in proportion to the distance driven. The first two drawings (Fig. 3.4) are complementary front views of the route indicator: a view with the cover of the casing in place (left) shows the strip map exposed beneath a glass-covered opening, and a view without the cover (right) reveals the inner workings. The bottom right shows a worm gear connected by "a flexible shaft [to] the front axle or front wheel of an automobile"—essentially an odometer cable, although the patent says nothing specific about measuring distance traveled.

As with a present-day in-vehicle GPS navigator, the direction of travel is toward the top of the glass display window. As the strip map moves downward, angular bends show turns at intersections (Fig. 3.4, left). Note the dashed lines that graphically pull the route back to the center of the strip—akin to a GPS reorienting its display after the driver completes a turn. So that the length of any part of the strip accurately reflects the corresponding distance along the route, the indicator had to be adjusted manually to match the advance of the strip map (drawn at "two inches to the mile") to the vehicle's speed.

Two of the patent's six drawings (Fig. 3.5) describe schematically the route's transfer from a conventional road map (right) to a narrow strip map (left), which "does not conform to the actual curves of the route." Heavy lines show the route to be traveled, thinner lines reflect crossroads, and evenly spaced, labeled tick marks along the right edge of the strip show distance traveled in miles. The numbers 1 through 11 on both drawings represent corresponding points along the route.

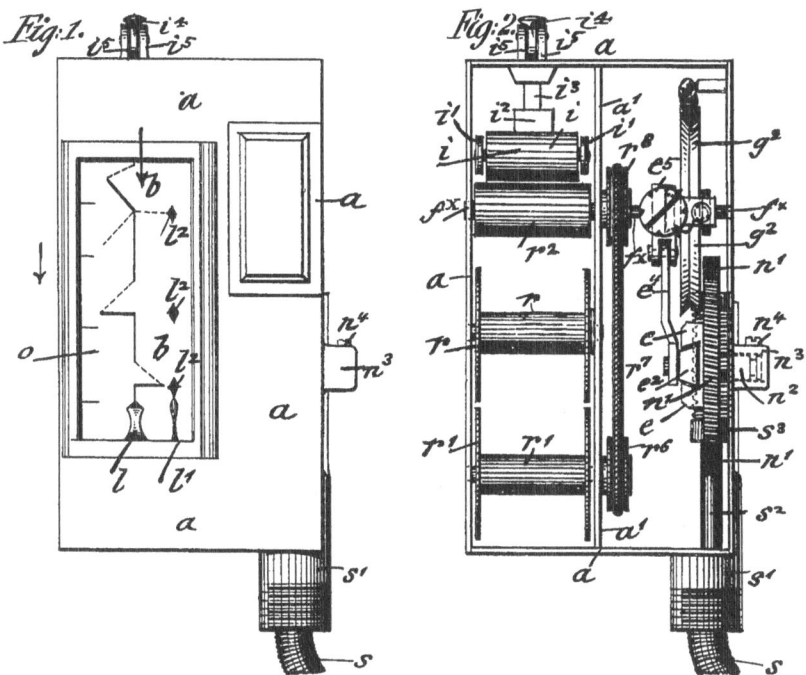

Fig. 3.4 Front view of the Lindenthaler and Protz's "Route-Indicator for Automobiles" with the cover in place (*left*) and removed (*right*) (US Patent 915,976; 1909)

Like a GPS navigator, the route indicator alerted the driver to upcoming events. Note the small holes punched into the strip at positions labeled i^2 (just outside the strip to the right) to mark "the more important intersections." At appropriate points along the route, a hole aligned with a contact-spring would close an electric circuit that rang a bell.

Their patent suggests the device was intended for a well-heeled automobile owner who wanted to micromanage his chauffeur. In addition to helping "an automobilist travel over comparatively unknown roads without the necessity of making inquiries on the road or mistakes," the route indicator would "facilitate the change of chauffeurs [insofar] as any chauffeur can readily drive the automobile over routes even if they are not specifically known to him." With knowledge of the road system no longer an essential skill, chauffeurs could be replaced more readily and paid less—an early warning of deskilling as a consequence of automation.

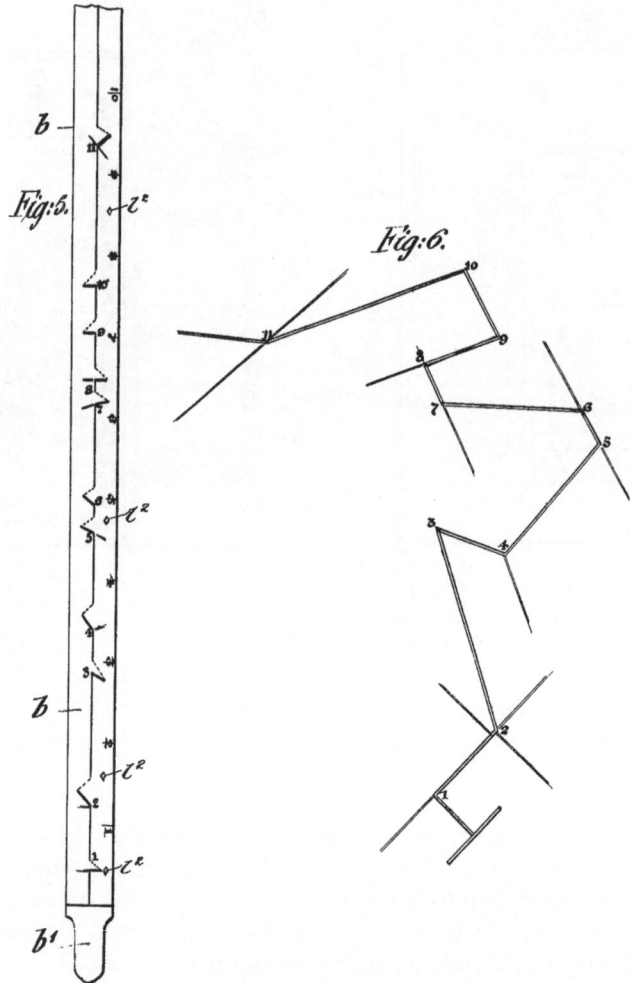

Fig. 3.5 Lindenthaler and Protz's strategy for converting a conventional route map (*right*) into a strip map (*left*) (US Patent 915,976; 1909)

To this end, an enthusiastic owner could "make a record of his own by taking a blank-strip of paper, placing it in the machine and marking the starting point and then the intersecting points at the right or left," and so forth. Anticipating increased automobile ownership, the inventors also recognized that "for specially favored routes covering greater distances, the strips may be published for the use of automobilists and sold at a small price." Even so, I found no evidence that Lindenthaler and Protz ever manufactured or licensed their invention.

Ingenious strategies mean little unless incorporated in the formal claims at the end of a published patent. Lindenthaler and Protz had an abundance of ideas, but not all of them met the Patent Office's standard for originality. Their original application reveals a naiveté that haunts inventors who fail to search for related foreign patents. All three of their claims were rejected by a conscientious patent examiner who discovered unacceptable similarities to a patent awarded in Britain in 1904 and two others awarded in France the following year.[20] In rejecting their third claim, the examiner also noted Johnson's 1880 patent, which had apparently anticipated parts of one of the French inventions. In Patent Office lingo, these similarities or overlapping claims are termed "interference."[21]

The inventors' attorneys no doubt played a key role in restructuring the patent to focus on the device's inner workings. When their revised application was approved, in March 1909, the three original claims had been reduced to two, and their introductory phrase "A route indicator" was now "*In* a route indicator" [italics mine]. Moreover, the revised claims focused on the "driving-roller," "shaft," "worm-wheel," "oscillating lever," "pivoted spring-actuated lever," and other elements inside the casing, not on the "indicator strip," now mentioned only once in just one of the two claims. A moving strip map was no longer patentable.

Originality was not the only issue. The patent examiner found numerous omissions and inaccuracies in their application, suggesting that the inventors' attorneys had failed to look carefully at the text and drawings before submitting the application.[22] For instance, although the reels are described verbally as "removable," the drawing had "no means for removably securing them." One discrepancy—"Either the bell and circuit referred to in lines 16 and 17, page 3, should be shown on the drawings, or the holes i^2 and contact spring i' should be omitted"—was addressed by adding a bell, battery, and wiring to a revised drawing (Fig. 3.6). And because "the flexible shaft s is only a part of the specified 'means',", they had violated the Patent Office's Rule 50, which required that every feature of the claims be shown in the drawings.[23] Substantially reworked claims introduced by "*In* a route indicator" addressed this objection.

In contrast to Lindenthaler and Protz, whose claims were purposely narrowed by the Patent Office to avoid interference with existing patents, Joseph W. Jones, of New York City, was able to patent an invention that

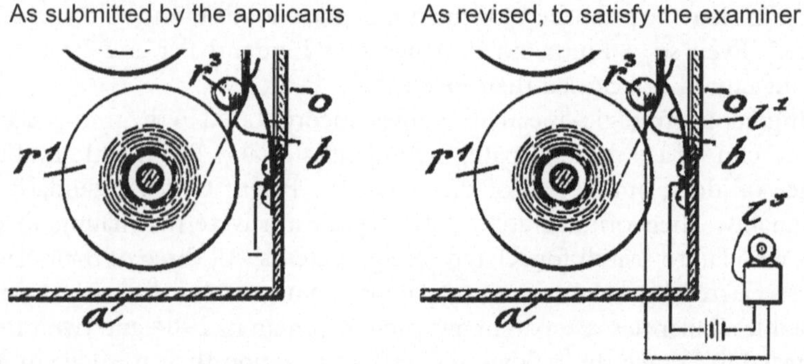

As submitted by the applicants   As revised, to satisfy the examiner

Fig. 3.6 Lindenthaler and Protz's application described an alert bell, which was missing from their original Fig. 3 (*left*). To address the patent examiner's complaint, they added a bell and electrical circuit to the revised drawing (*right*) (US Patent 915,976; 1909)

Fig. 3.7 Lead drawing in Jones's patent shows his route indicator connected to a front wheel by a flexible shaft (US Patent 1,040,345; 1912)

directly linked a front wheel to a hand-held indicator.[24] The first drawing (Fig. 3.7) of his aptly titled patent "Combined Road-Map and Odometer" shows an open car with a woman passenger in the back seat holding an indicator connected to the front axle by "flexible shafting … of sufficient length to reach any part of the vehicle." The only non-obvious part of the drawing is a hook (item 1 in Fig. 3.7) for hanging the indicator on the dashboard. The man in the front seat, reaching backward toward the passenger, is wearing goggles and a chauffeur's cap, and his employer or significant other, a well-dressed female wearing a fashionable chapeau, might be the prototypical backseat driver.

Jones's patent, filed in August 1909 and issued in October 1912, over three years later, was not his first. A mechanical engineer and entrepreneur born in Saratoga, New York, in 1876, Jones received more than 30 patents in a career that spanned more than half a century.[25] His inventions include a transformative means for recording sound, an electric automobile horn, a speedometer, a decorative tire tread, a taximeter, an aircraft tachometer, an electrically powered "massage apparatus," and a rotary device for polishing shoes. In 1896 he held a summer job in Washington, DC, as a laboratory lackey for sound recording pioneer Emile Berliner, whose disk record phonograph would soon eclipse Thomas Edison's cylindrical phonograph. Jones saw a better way to engrave record grooves on a wax master, and the following year he filed a patent claim, approved in 1901, for an improvement that transformed the disk phonograph industry after he sold it to the Columbia Graphophone Company for $25,000—his seed money for becoming an independent inventor.[26] Jones was largely self-educated, according to coding sheets for the 1940 Census, which list him as having completed only eight years of schooling.

Jones's route indicator reflects his early experience with the patent system, a strong interest in automobiles and precise measurement, and his invention of a speedometer manufactured by his own company—making things was as much an ambition as inventing them.[27] He began experimenting with devices for measuring speed in 1897, when he was 21, and three years later he founded the Jones Instrument Company.[28] Automotive historians eager for a famous first point to the "Speed-O-Meter" he installed in a Winton motorcar entered in a 1901 endurance run from New York to Buffalo.[29] Two years later he applied for a patent, which was issued in 1904.[30]

As a classic example of one invention inspiring another, the Jones Speedometer and his "Combined Road-Map and Odometer" each have a flexible shaft (speedometer cable) and a gear-driven attachment to a front wheel. Moreover, the flat, round shape of his road map is akin to a phonograph record—his patent even calls it a "turn-table." Unlike a conventional odometer, which registers accumulated mileage as a series of aligned digits on rotating coaxial drums, the Jones odometer displayed

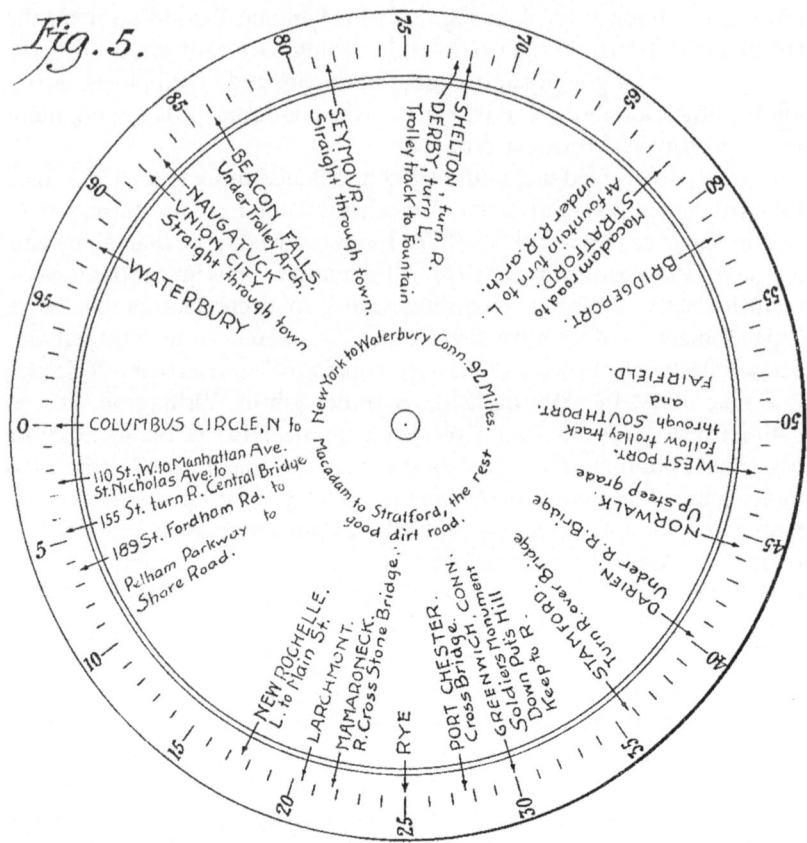

Fig. 3.8 A disk describing the route from Columbus Circle in Manhattan to Waterbury, Connecticut. Although this drawing is the fifth of five drawings spread over two pages at the beginning of the published patent, it was sufficiently important in describing the invention to share the first page with Fig. 1 (US Patent 1,040,345; 1912)

mileage graphically on the periphery of an interchangeable disk on which tick marks spaced a mile apart supplement numbered tick marks every fifth mile (Fig. 3.8). A set of gears inside the casing reduced the rapid rotation of the wheels to the slower, more precise counterclockwise motion of the disk, which turned as the vehicle advanced, and aligned the current distance traveled with a pointer at the far right (not shown on the patent drawing). In addition to mileage, the disk described road conditions, indicated turns, and flagged bridges, railroad crossings, road intersections, and other landmarks useful in following the route. In this example, the motorist would turn left onto Main St. in New Rochelle, 17.5 miles from the starting point, at Columbus Circle. Text near the center of the disk described the length of the route and its condition: in this case, 92 miles to Waterbury, Connecticut, and paved with macadam as far as Stamford and "good dirt road" beyond.

A decade of experience with the Patent Office did not spare Jones the frustration of repeated rejection of his patent claims. His application, filed in early August 1909, through a Washington, DC, law firm, was promptly rejected in late September because of similarity with two patents that "show the use of a flexible connection from the wheel of the vehicle to the odometer," and another that "shows a dial with a distance scale and a chart giving the names of places at corresponding distances."[31] Although none of these inventions was a route indicator, all involved components that argued for narrowing Jones's claims.[32] Amendments were made in response to each rejection, but the examiner continued to find snags. Altogether, Jones's application was rejected five times because of either overly broad claims or conflict with seven other patents, including that of Lindenthaler and Protz, whose rights included "mak[ing] this device portable and the drive shaft thereof of any desired length."[33]

A personal touch seems to have helped. In early September 1910, a week after the third rejection, an "Entry of Appearance" was added to the record when three attorneys from Mauro, Cameron, Lewis & Massie, the law firm representing Jones, visited the Patent Office to deliver a new response, which not only challenged several of the examiner's points but also called the rejection of one of the claims "clearly untenable."[34] Nonetheless, the vetting went through two more cycles of rejection and amendment until early May 1911, when a letter from the Patent Office conceded that the single remaining disputed claim "on further consideration … appears to be allowable without further amendment."[35]

Although Jones's claims of originality had been significantly altered to meet the examiner's objections, and reduced in number from nine to five, most of the wording that preceded the claims remained intact, including a clear statement of the invention's broad intent "to enable the automobilist to ascertain how far along on his journey he is, and to keep him posted generally as to the route; to enable him to know how many miles he has already traveled and how many more remain to be traveled to reach his destination (or any other point on the route); to enable any occupant of the vehicle, even on a rear seat, to obtain the same information; and to enable the automobilist to make his own chart as he goes over a new route, or to substitute a new chart when desired." The turn-table could be thrown in or out of gear so that the user could swap in a new disk for a different journey or join an existing route at an intermediate point.

There was one more hitch. Although the examiner's words "appears to be allowable" hinted at imminent approval, his next paragraph reported "a probable interference" with another, unnamed patent. Three weeks later, a letter from the Patent Office's Interference Division described several points of similarity between Jones's invention and a "Route Indicator" devised by Jay B. Rhodes, of Kalamazoo, Michigan.[36] "The question of priority," Jones was informed, "will be determined in conformity with the Rules."

Despite some similarities in gearing, drive train, and rotation, the interference was insufficient to undermine Jones's application. Rhodes had invented an odometer-like device that gave trip directions, but he used a series of coaxial cylinders (as discussed later), rather than a disk, and the two inventions looked markedly different.[37] Moreover, Rhodes had filed his application in late July 1910, nearly a year after Jones, who obviously had priority of invention for whatever narrowed claims the patent examiner was willing to accept.

Even so, Rhodes's application was pronounced "examined and allowed" less than two months before Jones and his attorneys learned of the interference.[38] Although the case files for Jones and Rhodes are silent on whatever deliberations ensued, more than a half year later, in early February 1912, the Patent Office's "Examiner of Interferences" awarded Jones a "favorable" decision. A month later his application was pronounced "examined and allowed," and after fees were paid and loose ends resolved, Jones received his patent in early October 1912, a year after Rhodes.[39]

Like many inventors with a compelling idea, Jones began to market his invention well before his patent was approved. He called the interchangeable disks "Live-Maps," the mechanism that advanced them the "Jones Live-Map Meter," and the overall system the "Jones Live-Map." Manufactured by the Jones Live-Map Meter Company, the device was sold through United Manufacturers, a New York marketing co-operative, which placed a full-page ad in the 19 May 1910 issue of *Life* magazine.[40] The upper half of the page highlighted the Live-Map with three halftone photos and forceful copy ("emancipates you from slavery to great, flopping maps and profound route-books that you can't make head or tail of without stopping"), while the lower half touted two other Jones products: the Jones Speedometer ("geared to the truth") and the Jones Electric Yobel, an "urgent, yet civil" horn with "a crisp, snappy signal" that "secures quick co-operation and instant right of way."

Bold publicity challenged conventional road maps. A short article published three months earlier in *Cycle and Automobile Trade Journal* read like a press release.[41] It was "impossible to get lost with this device in operation," readers were told, and the "circular route cards [indicate] the best route from one city to another." Moreover, "the advantages of handling a compact device in the wind will readily be seen by those who have endeavored to handle a map or even a book … in a fast car." Invented by "J. W. Jones, the well-known speedometer man," the Live-Map Meter had been exhibited at the Grand Central Palace Show and sold for $75. Twelve of the paper disks were included, and additional disks could be purchased for 25 cents, or "in quantity" for 15 cents. The device was about nine

Fig. 3.9 Detail of portion of a Live-Map disk (adapted from image provided by Wikipedia Commons and Museum of Science and Industry, Chicago, Illinois)

inches in diameter, and a glass cover protected the eight-inch disks from rain.[42] As Fig. 3.9 illustrates, the printed disks, with tick marks every fifth mile and every mile numbered, were more detailed than the patent's art-work implied.

Hyped publicity compared the disks to a record collection. A February 1910 *Saturday Evening Post* ad identified Jones as the "inventor of the disc phonograph record" and proclaimed the Live-Map "the phonograph of the road."[43] Although asserting that "the disc records cover[ed] the entire world" was blatant overreach, a catalog of 500 disks for North America gave the claim some credence.[44] The sales pitch promised ease of use ("You insert the record of the trip you want to make [and] the Live-Map 'plays' it.") and exceptional expertise ("To have it with you is like having in your car a man who knows every road, every corner, every crossing, every landmark, every puzzling fork and cross-road in the whole world."). Readers were encouraged to write in for *The Live-Map*, a "luxurious free book that tells you all about it."

Those who did received a 32-page booklet with posed photographs, a long and ominous subtitle (*The Instrument Which Knows Every Road, Every Corner, Every Puzzling Fork and Cross-Road in the Whole World, Described and Pictured Together With the Grim Story in 19 Photographs, of What Happens Without It*), and tragic vignettes with fairy-tale characters crafted to heighten fears of getting lost.[45] The first was "the Evil Genius of the Road," who deliberately sends a motorist asking directions "the wrong way." The next was "the Genius of Misinformation," who "meant to be obliging [but] sent us seven miles around through swamps and over mountains." There were also the rural trolley tracks marked only by a rock painted "Get right with God," the stammering man "who delayed us and upset us," and the "old lady [who] assured us none of the roads there-abouts would take us to Titusville." The contrast between geographically challenged rubes and wealthy male motorists wearing fur coats enhance the booklet's value among collectors of automobilia.

Jones was not only an inventor and entrepreneur but also a well-connected advocate for improved roads. In August 1910 the *New York Times* reported his return from three months in Europe as a US delegate to the International Road Congress, in Brussels.[46] The conference under-scored the need for standardized construction and proper maintenance. The article identified Jones as a director of the Touring Club of America, which compiled and distributed the Live-Map disks.[47] An October 1910 *Times* article reported he "has a large force of men charting out the good roads and main highways of this country ... for distribution through the Touring Club of America."[48] Jones himself was part of that force: on vacation in the South, he had recently checked out a newly improved highway from Jacksonville to Tampa, and was "on his way north in his automobile, laying out a new New [sic] Capital Highway trail to this city."

Jones never patented an improvement to his Live-Map Meter, but his younger brother, Ernest Albert Jones, devised a way to compensate for a driver's inability to follow precisely the centerline of a prescribed route. Because meandering about the intended route lengthened the actual distance traveled, the route display tended to over-run the vehicle. To lessen the need to reset the vehicle's position along the route, a "compensating device" interrupted the flexible shaft running from the wheel to the indicator with a set of gears, springs, and friction disks (Fig. 3.10, right) "coupled to the steering mechanism."[49] Although the details are obscured in a fog of technical language, irregular steering reduced the rotational motion of the wheel transmitted up the flexible shaft, thereby retarding the advance of the route map. Because the invention could be used with other route indicators, Ernest's patent, titled merely "Route-Indicator Mechanism,"

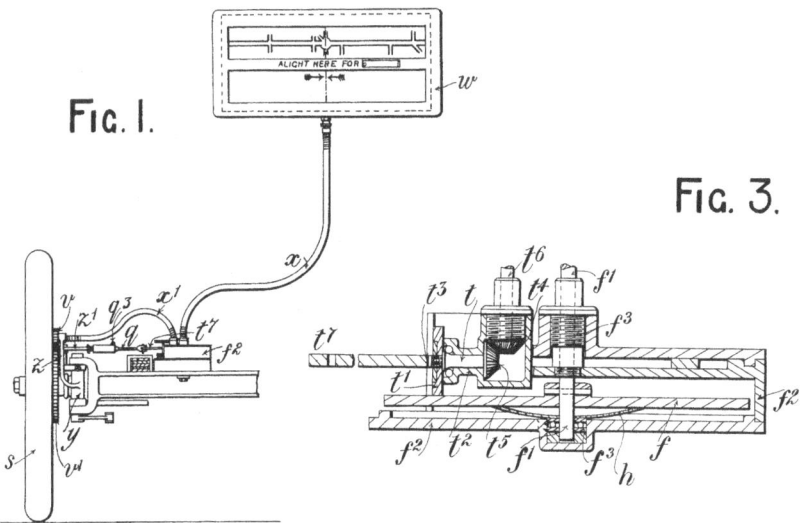

Fig. 3.10 Two of the five drawings for Ernest Albert Jones's "compensating device" show a mechanism with two friction disks inserted between the wheel and the route indicator (US Patent 1,092,147; 1914)

does not mention Joseph's Live-Map Meter, and the only map in its draw-ings (Fig. 3.10, left) was a strip map, not a circular Live-Map. No point in lessening the patent's value with needlessly narrow claims.

In contrast to his brother's experience, Ernest met minimal resistance at the Patent Office. He applied in late July 1913, and received official approval eight months later, after several requests to amend wording and alter one of the drawings. There were no rejections, and the examiner's initial letter conceded, "The claims all appear directed to patentable mat-ter."[50] The most prominent editorial change was a request, late in the correspondence, to abbreviate the "unnecessarily long" title "Route Indicators or the Like for Public Service Vehicles."[51] Ernest apparently believed his most likely buyer was a motorbus operator or car service, not an individual motorist, which suggests the Jones Live-Map was not meeting sales expectations. Nonetheless, he was sufficiently optimistic to register patents in Canada, Great Britain, France, Germany, Switzerland, and the Union of South Africa.[52]

Ernest was more cosmopolitan than his brother. His patent described him as a United States citizen "and a resident of New York but temporar-ily residing at Hampstead, London, England." Four years younger than Joseph, he too had been born in Saratoga, New York, and also seems to have lacked formal training as a mechanical engineer.[53] He was not picked up in the 1940 Census, which inquired about the highest grade of school completed, because he had been living off and on in England since around 1903, according to passport applications and native-citizen registrations.[54] His reasons for living in England involved motor vehicles or employment as a consulting engineer. On a 1921 "Affidavit to Explain Protracted Foreign Residence," he claimed to have come to Britain as "special repre-sentative of J.W. Jones, of the late Jones Speedometer Indicator Co. … and P.S. Jones, Attorney at Law" (another brother, also based in New York), and to have worked "during the war … as [a] consulting engineer on the manufacture of munitions." He apparently remained in Britain for the rest of his life, but I have yet to find a date or place of death.

Ernest's short involvement with the patent system focused on the auto-mobile. He patented three additional inventions, all in England. The last, patented in 1915, was for an improved "kinematograph target apparatus," a firing-range device that projected a target on a screen and halted the pro-jector when the bullet struck the target.[55] The others were an improved "Carburetting Apparatus for Internal Combustion Engines" and a route indicator that displayed advertising.[56]

By contrast, his brother Joseph continued to invent after 1913, when he sold the speedometer business to Johns-Manville, a manufacturing conglomerate best known for its asbestos insulation.[57] In 1915 he founded Jones-Motrola, Inc., which marketed an electrical motor that replaced the hand crank on the wind-up phonograph—the July 1919 *Popular Engineer* called it "one of the latest and most successful additions to the talking machine industry."[58] In 1924 he started the Jones Radio Manufacturing Company, but got out of the business around 1930, when he was nearing age 65.[59] His last patent, filed in 1943, was an aircraft gunnery training aid that simulated the appearance of an airplane as seen from various distances through a gun sight.[60]

A would-be competitor in the automated road-map business was Jay B. Rhodes, mentioned earlier because of interference between his patent and Jones's. Rhodes's very different design surrounded a brass cylinder with alternating metal bands representing distance and direction (Fig. 3.11). A route was programmed by consulting a "route book in which directions are given for stations" or turning points, and by setting the bands to describe distance and direction to the next station.[61] Distance bands were graduated in miles and tenths of miles and could specify segments as long as five miles. Direction bands used letter codes like L (for turn left), RF (for take the right fork), BR (for bear right), and S (for straight ahead).

The display was not easy to read. Specific directions for a point along the route were highlighted by a movable "index" (60) with a pair of vertical arms (63) and a "pointer" (62), as shown in the lower part of Fig. 3.11. At this point along the route, a right fork is 3.4 miles ahead. Look closely for the R over F, meaning right fork, and note that the short horizontal line between the R and F is aligned with a tick mark on the mileage band immediately to the right indicating that the fork in question is 3.4 miles down the road. Because the band is rotating upward, the three should shortly rotate into view from below.

Patent drawings also show an attached speedometer and a bell that rang as the motorist approached a station: a crude, inaudible forerunner of the talking GPS navigator. Although Rhodes's "cylindrical chart" merited four paragraphs in a 1911 *Scientific American* article on route indicators, I doubt his "hand-adjusted guide" was ever marketed, at least not effectively.[62] A snapshot captioned "A hand-adjusted guide: method of resetting directions and mileage rings" shows a working prototype, but

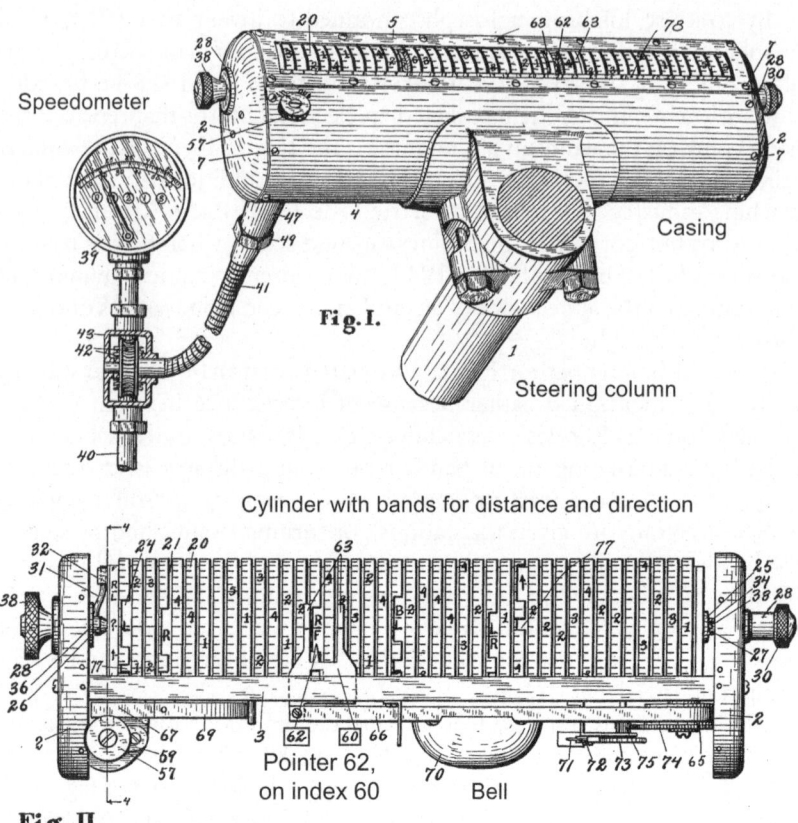

Fig. 3.11 The first two of the 13 figures describing Jay B. Rhodes's programmable route indicator. Annotations include labels highlighting the index and pointer alerting the driver to a right fork 3.4 miles ahead (US Patent 1,005,474; 1911)

extensive searching failed to locate a single advertisement or news article focused on what the article (but not the patent) labeled the "Pathfinder."

Similarly unremarkable is the patent case file in the National Archives. Rhodes asserted 39 claims, which were reduced to 25 in seven months that included three rounds of rejection and amendment. In his initial response, the patent examiner pronounced several of the claims redundant or not patentable, and rejected others because of existing patents, including Rhode's own patent, issued six months earlier for a less complicated, conceptually similar (and identically titled) "Route Indicator."[63] Early filings to establish priority can interfere with later refinements.

Rhodes fits the pattern of a prolific self-educated inventor. He was born in 1865 in rural Michigan, not far from Kalamazoo, where he lived most of his life. In 1911, two months after he received his Pathfinder patent, his brother Bert patented a "Route Indicator" that enhanced Jay's apparatus with a single broad cylinder that displayed "the name of the place ... the vehicle is passing" as well as distance and direction to the next station.[64] Because the names apparently had to be composed letter by letter with movable type, the patent probably never yielded a working prototype. Bert, who was four years younger than Jay, lived until 1943—long enough for the 1940 Census to note his having left school after grade six.[65] Jay, who died in 1931, probably had a similarly short formal education. The 1900 Census found the brothers married, living nearby outside Chicago, and working as machinists. Jay moved back to Kalamazoo in 1903, and subsequent enumerations reported his occupation as "manufacturer [of] automobile sundries" (1910), "consulting engineer" (1920), and "inventor" (1930). Local historians who have dubbed him "Kalamazoo's Patent King" cite successes like a widely used dispensing can for motor oil and a safety razor he manufactured and marketed himself, along with Rhodes Blades.[66] His Pathfinder was one of many bright ideas, clever but complicated, and of diminishing usefulness as signposting improved.

A competing device that also compensated for inadequate signposting was invented and manufactured by Lee Sherman Chadwick, of Pottstown, Pennsylvania, about 40 miles northwest of Philadelphia, and patented in 1911 and 1916.[67] As the vehicle approached an intersection or other danger point, the Chadwick Automatic Road Guide rang a gong or electric buzzer and simultaneously displayed a large image of one of the ten icons in Fig. 3.12. Locations of these events along the route were recorded on a disk about five inches in diameter, made of aluminum or thin, stiff paper.[68] The disk contained ten evenly spaced rings (Fig. 3.13), each representing a particular type of warning, and perforations along the ring marked locations at which the system would signal that particular type of turning point or danger. Like the Jones Live-Map, the device was connected to the wheel through a speedometer cable so that the disk turned slowly as the vehicle advanced.

Chadwick was probably inspired by a home entertainment system popular early in the century: the self-playing piano, or player piano, which relied on perforated rolls of pre-recorded music and compressed air

	Straight through town, or take center road		Danger: bridge ahead, or use perfect control
	Crooked road ahead, use caution		Go slow: speed trap, or danger of arrest
	Rough, bad road, ahead, use caution		Railroad crossing, or mulitple cross streets
	Sharp left ahead, or take left-hand road		Curve to right, or take right-hand fork
	Curve to left, or take left-hand fork		Sharp right ahead, or take right-hand road

Fig. 3.12 Signal icons displayed by the Chadwick Automatic Road Guide. Compiled by author from text and image in Chadwick's 1911 patent (US Patent 1,002,368; 1911)

pumped with foot pedals.[69] In the same way that tiny holes in a music roll were aligned in positions representing particular notes on the keyboard, perforations in the rings of the Chadwick Automatic Road Guide were matched with thin tubes, one for each of the ten warnings. When a hole in the disk aligned with the appropriate tube, a surge of compressed air rang the bell and activated the corresponding visual signal. Although Chadwick's patent described rods, rather than tubes for compressed air, a 1912 *Motor Age* article that discussed a working prototype mentioned only the pneumatic trigger.[70] The patent was sufficiently broad to cover the modification.

Chadwick originally applied for a single patent covering mechanisms for recording a master disk as well as for playing a recorded route, but the Patent Office considered them different inventions, which had to be "examined in different divisions of the Office."[71] The application was divided accordingly, and Chadwick eventually received two patents. Oddly, the recording patent was approved in only 18 months, whereas the signaling patent took over six years. Its ten letters of rejection and 16 replies proposing amendments suggest that the Chadwick Route Guide was very much a work in progress.[72]

Marketing must have been a struggle. Two models of the Route Guide were advertised, for $55 and $75, but the Jones Live-Map Meter, with

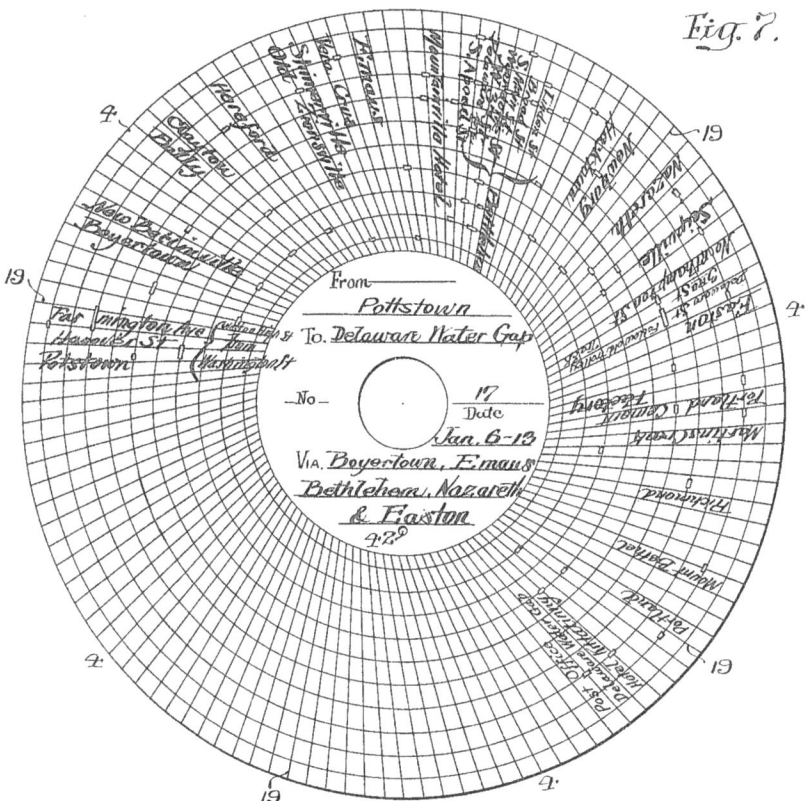

Fig. 3.13 Annotated example of the disk record used with the Chadwick Automatic Road Guide. Rings correspond to the ten different signals in Fig. 3.12, perforations indicate signaling location, and transverse lines represent position along the route (US Patent 1,180,239; 1916)

a $75 price tag, was a strong competitor even though it lacked a warning bell.[73] Jones had the support of the Touring Club of America, which compiled and sold his Live-Map disks, whereas Chadwick was apparently his own content provider. I found no catalog for Road Guide disks and am skeptical of *Motor Age*'s reports that "road records furnished at this time cover most of the eastern states" and that "it is intended within a few months to cover the more prominent routes all over the country with the automatic records."[74]

Unlike Rhodes, the Jones brothers, and the other inventors discussed above, Chadwick had a college degree: a B.S. in mechanical engineering from Purdue University, class of 1899. Born in 1875, the son of a Vermont farmer, he was fascinated with automotive horsepower, particularly speed trials up mountain roads. In 1903 he began manufacturing his own cars, first with four-cylinder and then with six-cylinder engines. One model, the Great Chadwick Six, was capable of 100 mph and won several speed competitions. He sold the company in 1911, to focus on the Road Guide, which was manufactured at a Cleveland, Ohio, foundry where he had been hired as chief engineer. The owner valued his talent and let him set up a small production line. As the road indicator market dried up, Chadwick took on greater responsibilities at the metal products company and eventually became the firm's president.[75]

As the Theory of Multiples would predict, other inventors also saw a need to automate road maps for easy reading. Open cars, poor roads, and unreliable signage created an opportunity for creativity reflected in multiple patents issued between 1910 and 1920 for route indicators—a multiplicity that reflects diverse ways of moving a strip map through a viewing window and alerting the motorist to turns and hazards. Because patentability refers to the means rather than the end, a new and arguably non-obvious improvement could be patentable even though it might be less efficient that other ways of achieving the same result. The patent system was always content to let the market sort out practicality and profitability. That most of these ideas came from inventors new to the patent system reflects a quirky mix of perceived need, cleverness, self-confidence, and a thirst for fame or fortune.

Most of these improvements were simpler than the comparatively sophisticated inventions of Jones and Chadwick. For instance, the "Route Indicator for Vehicles" patented in 1910 by Frank Feilhuber, a Newark, New Jersey, railroad brakeman, was just a flexible strip map stretched between two reels and pressed against a "transparent plate."[76] Clearly intended for manual operation, it was not conceptually different from Harry McCafferty's handlebar-mounted map holder (Fig. 3.1), patented more than a decade earlier. By contrast, the "Mechanically Operated Road Map" patented in 1913 by New Rochelle, New York, tobacco broker and accountant Max Bremsy was (as its title implies) "connected preferably by a flexible shaft to one of the wheels."[77] A conceptually similar two-reel device, it included two additional rollers to hold its flexible transparent map taunt and a light bulb below for backlighting at night. Unlike Feilhuber,

Bremsy held another patent, for a "Clothes-Line Clamp," issued the same year.[78] According to a 1914 list of licensed vehicles, Bremsy owned an Overland automobile, priced well below the Great Chadwick Six.[79]

Although Manhattan resident George Boyden might not have owned a car, he was surely familiar with driving and wayfinding insofar as coding forms for the 1910 and 1920 Censuses list his occupation as chauffeur for a "private family." In 1914 he patented a "Chart for Vehicles," a reel-to-reel device with similar "idle rollers" for holding taunt the flexible strip map, or "web."[80] His patent included 15 claims, showed a "flexible shaft" connected to a wheel, and addressed recording as well as following routes. Boyden held another patent, issued two years later, for a "Vehicle Signaling System," which "provide[d] a means of announcing to the driver … the directions for following a predetermined route."[81] *Announcing* was not idle rhetoric—Boyden's apparatus device included a phonograph, and one of the drawings showed a "megaphone" on the dashboard directly in front of the driver (Fig. 3.14). As far as I can tell, neither patent was ever developed.

Another part-time inventor of route indicators was George Deardroff, of Occoquan, Virginia. Born around 1871, he completed two years of high school and worked at various times as a railroad agent, brickyard superintendent, and poultry farmer. All four of his patents were based on "webs" that ran over or between multiple rollers on their way from a supply reel to a take-up reel. His simplest design, a "Route Indicator" patented in 1913,

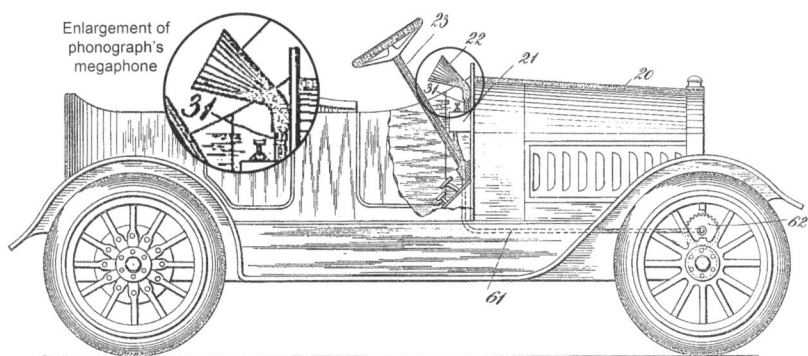

Fig. 3.14 George Boyden's "Vehicle Signaling System" included a phonograph and played an audio recording of directions and warnings through a megaphone in front of the steering column (US Patent 1,168,053; 1916)

displayed a simple strip map.[82] By contrast, his "Road Map for Automobiles," patented the same year, used different gearing to display a sequence of verbal travel directions.[83] But his "Route Indicator" patented in 1917 not only displayed a list of directions but also included a speedometer and separate odometers labeled "season" and "trip."[84] All were connected to a wheel by a flexible cable, in contrast to his "Road Map" patented in 1916, which included a backlighting lamp and used electrical impulses to synchronize its web to a wheel.[85] I doubt that any of his patents were developed.

Foreign applicants who used the US patent system included Pio Papini, a resident of Florence, Italy, who received an American patent in 1914 for an "Indicator for Illustrating and Signaling the Route of a Vehicle."[86] He called his strip map a "band," added sprocket holes along the edges, linked its motion to "a train of speed reducing gearing," and used appropriately placed holes to trigger audible and visual alerts to turn right, turn left, and slow down. A fourth alert, intended for passengers as well as the driver, called attention to the "particulars of the road."

Among the simpler plans for route indicators were the drawings with which Phoenix, Arizona, resident Celora Stoddard described a reel-to-reel device that displayed a sequential list of directions consisting of words, symbols, or numbers.[87] Patented in 1916, it was positioned next to the odometer, to which it need not be connected. (A large "milled knob" allowed for manual operation.) Stoddard had two years of college and ran his own investment company. Twenty-seven years old when he applied, he held no other patents and presumably did not develop this one.

Chicago resident William Reilly, who patented another route indicator that same year, also displayed sequential directions on a "tape," but used a flexible shaft to harness its movement to a wheel.[88] Reilly, who was a machinist at a glue factory, might have devised a working prototype, but I found no evidence he developed the invention. This was his only patent, filed when he was about 22 years old.

Two route indicators were patented in 1917. Howard Cranmer, of Wichita Falls, Texas, connected his device to a wheel by a "power transmitting shaft," displayed a sequence of verbal and symbolic directions on a "tape," and provided backlighting and an audible alert with a buzzer triggered by holes in the tape.[89] A former carpenter who worked in the lumber business, Cranmer was in his early 50s when he filed his patent. He had only this one patent, which he seems not to have developed.

By contrast, Henry Hubschmidt, a Passaic, New Jersey, resident, was a 31-year-old business manager at a private school at the time he filed. His more sophisticated patent describes a "ribbon map" with text on the left side and simple graphic symbols on the right. It included a battery, an illuminating light, and a bell or buzzer.[90] With both an odometer (including a trip counter) and a speedometer, it could either replace an existing speedometer or connect to a wheel through its own flexible shaft. Despite four years of high school and business experience, Hubschmitt apparently never developed the patent. He held two others, awarded in 1919 and 1925, respectively, for school furniture and a roof bracket.[91]

With the number of arguably original features drastically narrowed by the foregoing patents, it is hardly surprising that 9 of the 12 claims allowed in the "Route and Station Indicating Means" patented in 1918 by William Brien and Marvin Whittaker began with the phrase "In a route indicator, the combination."[92] Each of these nine described a configuration or modification of "a pair of spools," "winding gear," and "movable tape having perforations" sufficient to meet the Patent Office's standards for usefulness, originality, and non-obviousness, while the other three passed muster by injecting terms like "housing" or "hinged section." The accompanying drawings focused on the invention's complex interior, which supported a digital speedometer (non-electric of course), separate "seasonal mileage" and "trip" odometers, an alarm bell, a connection to a flexible shaft, and a tape showing a sequence of verbal directions and warnings. As an outlet for creative energy, the mechanical route indicator had already crossed the border between youth and maturity.

Both inventors lived in Indianapolis, but the basis for their collaboration is obscure. They were born in Upstate New York, about a hundred miles and 12 years apart. When they filed their patent application, Whittaker was about 56 and a manager at International Harvester, one of the city's largest employers, and Brien was a 44-year-old salesman, who later worked as secretary for the local YMCA.[93] They never developed the patent and held no others.

A refreshingly original idea emerged in 1920, when Edward Siegel, a 36-year-old shipyard plumber in Elmhurst, New York (within the Borough of Queens), received a patent for a "Vehicle Route Indicator Device" that advanced its list of driving directions in spurts, rather than gradually.[94] To do this, Siegel added a second representation of the route on a "flexible conductor" synchronized to the movement of the vehicle. His patent described this enhancement as "a relatively slow continuously movable member" integrated with "a relatively fast intermittently moving member" carrying the list of directions, and "an electromagnetic means for controlling the movement of the latter." A side-view drawing (Fig. 3.15) showed the tape with sequenced directions in the upper third of the casing, beneath a transparent viewing window and running from a supply reel (on the left) to a take-up reel (on the right). Farther down, a flexible conductor—a wire mostly insulated but uncovered in places—ran from a supply spool (on the left) to a take-up spool (on the right). Along the way the wire passed between a pair of metal rollers, arranged so that an uninsulated section would close a circuit that activated the motor that advanced the tape to display the next direction. The bare sections of wire represented locations along the route at which the device was to show a new warning or instruction.

Clever perhaps but neither practicable nor profitable. This was Siegel's only patent, and like most inventors of improved route indicators, he never developed it. Although a few other patents were issued in the early 1920s, automobile buyers had little use for expensive mechanical route indicators

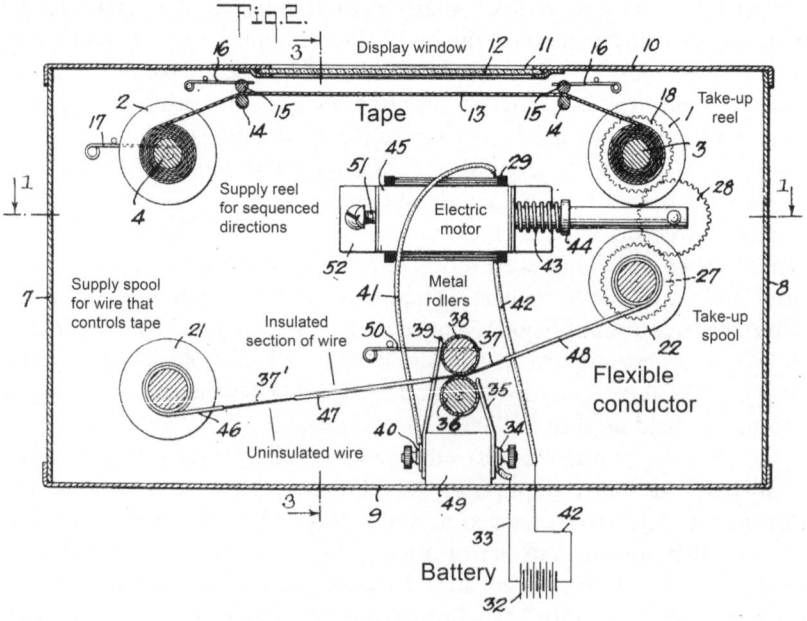

Fig. 3.15 Side view of Edward Siegel's strategy for synchronizing a tape with sequenced route directions to a flexible conductor linked to a motor that controlled the tape's advance (US Patent 1,350,244; 1920)

that favored a relatively small number of select routes and required frequent resetting. Paved streets, numbered highways, and reliable signage lessened fears of getting lost, and the closed car made it easy to use the growing number of inexpensive, generally reliable, and occasionally clever printed road maps and atlases.[95] The automated road map remained a dormant technology for more than half a century, until engineers began to experiment with electronic approaches.[96] As the odometer had been a catalyst for the mechanical route indicator, the integration of satellite positioning, the electronic street map, and miniaturized computers led to the commercially successful in-vehicle navigation system—an innovation that, in concert with situation awareness technologies, now threatens to take over the wheel.

Although the odometer was a catalyst, perceived usefulness was no less important in inspiring full-time inventors like Rhodes and a larger cohort of spare-time inventors like Boyden and Siegel, for whom the patent system was a means for proclaiming creative prowess, much like the parallel academic-scientific-technical literature. What's different is these amateur inventors' apparently limited knowledge of what their counterparts in diverse locations were doing—knowledge informed less by the *Official Gazette*, I suspect, than by rejection letters from patent examiners eager to point out similarities. Lust for fame or profit was no doubt encouraged by patent attorneys eager for business as well as by announcements of commercial offerings in periodicals like *Cycle and Automobile Trade Journal*.

What's intriguing about these early attempts to customize maps for route following is the inventors' clever generalization of cartographic symbols to fit the narrow format of a strip map and later by the route's reduction to a simple list of directions and the addition of visual or audible alerts to get the driver's attention. Though not immediately transformative, letting the map guide its user was a significant breakthrough.

NOTES

1. O. A. W. Dilke, "Itineraries and Geographical Maps in the Early and Late Roman Empires," in *Cartography in Prehistoric, Ancient, and Medieval Europe and the Mediterranean* (Vol. 1 of *The History of Cartography*), ed. J. B. Harley and David Woodward, 234–57, esp. 238–42 (Chicago: University of Chicago Press, 1987). For a broader discussion of the genre, see Alan M. MacEachren, "A Linear View of the World: Strip Maps as a Unique Form of Cartographic Representation," *American Cartographer* 13 (1986): 7–25.

2. "Ribbon map of the Father of Waters" (St. Louis, MO: Coloney & Fairchild, 1866). The map was rolled onto a spool and protected by wooden case. A digital copy is online at the Library of Congress, http://www.loc.gov/item/86691596/.

3. See, for example, League of American Wheelmen, *Road Book of the Michigan Division, League of American Wheelmen*, 4th ed. (Detroit: E.N. Hines, 1897). For an overview, see Douglas A. Yorke, Jr. and John Margolies, *Hitting the Road: The Art of the American Road Map* (San Francisco: Chronicle Books, 1996), esp. 13–18.

4. Christina E. Dando, "Riding the Wheel: Selling American Women Mobility and Geographic Knowledge," *ACME: An International E-Journal for Critical Geographies* 6 (2007): 174–210.

5. James R. Akerman, "Road Mapping in Canada and the United States," in *Cartography in the Twentieth Century* (Vol. 6 of *The History of Cartography*), ed. Mark Monmonier, 1339–50 (Chicago: University of Chicago Press, 2015).

6. James R. Akerman, "American Automobile Association," in *Cartography in the Twentieth Century* (Vol. 6 of *The History of Cartography*), ed. Mark Monmonier, 46–49 (Chicago: University of Chicago Press, 2015).

7. Straightening out a curving railroad made the route appear more direct, and labels perpendicular to the rail line allowed the mapmaker to identify more stations. See David Woodward, *The All-American Map: Wax Engraving and Its Influence on Cartography* (Chicago: University of Chicago Press, 1977), 31–36.

8. Harry C. McCafferty, "Map Holder," US Patent 605,969, filed 13 July 1897, and issued 21 June 1898. Ancestry Library Edition led me to the appropriate census taker's schedule for the 1900 Census, which also reported McCafferty's occupation, his birth in 1859 in Pennsylvania, and his marriage in 1885; a renter rather than a home owner, he could read and write.

9. John S. Kelso, Jr. and Peter L. Wilbur, "Holder for Road-Maps," US Patent 625,844, filed 8 November 1897, and issued 30 May 1899.

10. Ancestry Library Edition was a portal to the 1900 Census and the 1899 city directory, *Trow's General Directory of the Boroughs of Manhattan and Bronx, City of New York, for the Year Ending July 1, 1899*, 688, 1411.

11. John S, Kelso, Jr., "Incandescing Electric Lamp," US Patent 273,366, filed 11 October 1882, and issued 6 March 1883. Also see Thomas A. Edson, "Electric Lamp," US Patent 223,898, filed 6 November 1879, and issued 27 January 1880.

12. Harry W. Perry, "Some Remarkable Mechanical Road Guides," *Scientific American* 104.2 (14 January 1911): 33, 47–49.

13. Lincoln J. Carter, "Road Map," US Patent 770,350, filed 23 November 1903, and issued 20 September 1904.

14. Ancestry Library Edition was a portal to the 1900 Census and the *"Lakeside" City Directory for Chicago, 1900* (Chicago: Chicago City Directory Co., 1900), 385. Carter's other US patents and their numbers and dates of issue are "Fireproof Curtain" (828,532; 1906), "Stage Appliance for Theatres" (688,387; 1901), "Theatrical Appliance" (622,807; 1900), and "Theatrical Scenery" (a title used for three patents: [703,763; 1902], [747,045; 1903], and [843,583; 1907]).

15. "Biography of Carter, Lincoln J.," http://moviespictures.org/biography/Carter,_Lincoln_J.

16. For discussion of Chicago's preeminence in American commercial cartography, see Michael P. Conzen, ed., *Chicago Mapmakers: Essays on the Rise of the City's Map Trade* (Chicago: Chicago Historical Society for the Chicago Map Society, 1984), esp. 4–11.

17. William F. Johnson, "Automatic Station Indicator and Advertiser," US Patent 231,961, filed 31 December 1879, and issued 7 September 1880.

18. Frank J. Lindenthaler and John Protz, "Route-Indicator for Automobiles," US Patent 915,976, filed 30 November 1907, and issued 23 March 1909.

19. Ancestry Library Edition was a portal to many useful documents, in particular the hand-written census taker's schedules for the 1900, 1910, 1920, and 1930 federal censuses and for the 1915 and 1925 New York State censuses as well as Lindenthaler's World War I draft registration card, his record in the Social Security Death Index, and city directories of Manhattan and the Bronx for 1906 and 1908 and of Glens Falls, New York, for 1907.

20. G. A. Nixon (examiner) to Lindenthaler and Protz (c/o Goepel & Goepel [their attorneys]), 24 February 1908. The three patents are Bernhard Friedrich Sobotka, "Motor Car Road Indicator," British Patent 12,705, filed 5 June 1903, and accepted 4 June

1904; Henri Rousson, "Liseur de cartes déroulant automatiquement la carte de l'itineraire suivi, applicable aux automobiles" ["Reader to Automatically Unveil a Route Map, Applicable to Automobiles"], French Patent 348,260, filed 26 November 1904, and published 8 April 1905; and Charles Renac, "Dérouleur automatique de cartes routières" ["Automatic Dispenser of Route Maps"], French Patent 348,334, filed 28 November 1904, and published 10 April 1905.

21. For discussion of how patent examiners recognize and mitigate interference, see Fred K. Carr, *Patents Handbook: A Guide for Inventors and Researchers to Searching Patent Documents and Preparing and Making an Application* (Jefferson, NC: McFarland & Company, 1995), 105–12.

22. G. A. Nixon (examiner) to Lindenthaler and Protz (c/o Goepel & Goepel), 24 February 1908.

23. Albert H. Walker, *Text-Book of the Patent Laws of the United States of America*, 2nd ed. (New York: L. K. Strouse & Co., 1899), 94.

24. Joseph W. Jones, "Combined Road-Map and Odometer," US Patent 1,040,345, filed 6 August 1909, and issued 8 October 1912.

25. Although I have yet to find a complete obituary or biography, images of primary documents available through Ancestry Library Edition provide a few key details about Jones's life. According to his World War I draft registration, he was born on 7 June 1876, but his passport application says 27 June 1876—in his own handwriting in both cases. Ancestry reports Jones died on 7 March 1960, but provides no verification. That he lived into his 70s is consistent with other facts, including an application for a design patent filed in 1949 for a shoe-polishing machine; see www.google.com/patents/USD159264. His widely shared surname and no less common first and middle names (Joseph William) complicate the search for unambiguous details.

26. Alan Douglas, *Radio Manufacturers of the 1920s: Volume 2, Freed-Eisemann to Priess* (New York: Vestal Press, 1989), 77; Roland Gelatt, *The Fabulous Phonograph, 1877–1977*, 2nd ed. (New York: Macmillan Publishing Co., 1977), 132–33; and Rick Kennedy, *Jelly Roll, Bix, and Hoagy: Gennett Records and the Rise of America's Musical Grassroots* (Bloomington: Indiana University Press, 2013), 19–24. Also see Joseph W. Jones, "Production of Sound-Records,"

US Patent 688,739, filed 19 November 1897, and issued 10 December 1901.

27. Joseph W. Jones, "Speedometer," US Patent 765,841, filed 28 June 1903, and issued 26 July 1904.

28. Russ Furstnow, *The Antique Automobile Speedometer: A Study of Speedometer Entrepreneurs, Manufacturers and Early Speedometer Components from 1908–1927* (Flagstaff, Ariz.: Russ Furstnow, 2008), 2–3, 15–16.

29. The race was halted at Rochester after President McKinley was assassinated; see John A. Jakle and Keith A. Sculle, *Motoring: The Highway Experience in America* (Atlanta: University of Georgia Press, 2008), 11–12.

30. Joseph W. Jones, "Speedometer," US Patent 765,841, filed 28 June 1903, and issued 26 July 1904.

31. J. T. Newton (examiner) to Joseph W. Jones (c/o Mauro, Cameron, Lewis & Massie [his attorneys]), 21 September 1909.

32. See Lyman C. Perkins, "Odometer," US Patent 251,064, filed 27 August 1881, and issued 20 December 1881; John Davidson, "Cyclometer Watch," US Patent 622,884, filed 17 August 1897, and issued 11 April 1899; and Lake E. Fugate, "Register and Alarm," US Patent 708,632, filed 28 January 1902, and issued 9 September 1902.

33. Geo. P. Tucker (examiner) to J. W. Jones (c/o Mauro, Cameron, Lewis & Massie), 31 May 1910. Other rejections were dated 24 October 1910, 23 November 1910, and 7 March 1911.

34. Joseph W. Jones [through his attorneys] to Commissioner of Patents, 31 October 1910; and "Entry of Appearance," by S. T. Cameron, Reeve Lewis, and C. A. L. Massie, same date.

35. Geo. P. Tucker (examiner) to J. W. Jones (c/o Mauro, Cameron, Lewis & Massie), 6 May 1910.

36. Geo. P. Tucker (examiner) to Joseph W. Jones (c/o Mauro, Cameron, Lewis & Massie), 20 May 1910.

37. Jay B. Rhodes, "Route-Indicator," US Patent 1,005,474, filed 27 July 1910, and issued 10 October 1911.

38. Commissioner of Patents to Jay B. Rhodes (c/o Chappell and Earl [his attorneys]), 30 March 1911.

39. Interference Division, US Patent Office, decision on interference of Joseph W. Jones, "Combined Road Map & Odometer," with patent issued to Jay B. Rhodes, 5 February 1912; Commissioner of

Patents to Joseph W. Jones (c/o Mauro, Cameron, Lewis & Massie), 13 March 1912.

40. "Jones Live-Map" [advertisement], *Life* 55 (19 May 1910): 907.

41. "Jones' Live-Map," *Cycle and Automobile Trade Journal* 14.8 (1 February 1910): 221.

42. The aforementioned *Cycle and Automobile Trade Journal* article reported "the dial is about 8¾ inches in diameter," which is consistent with an announcement from the Skinner auction house that referred to the "interchangeable 8-inch dia. discs," and the "overall dia. [of] 9 in. [of the unit and twenty discs] in [its] original leather wallet." The device was sold for $764. See Skinner, Inc., "Jones Live-Map Meter" [auction date 29 July 2006], http://www.skinnerinc.com/2345/lots/37. By contrast, Christie's reported the sale of a similar named instrument "in [a] leather case—10in. (25.4cm.) wide" for only $143 at an 11 December 1996 auction; see Christie's, "An Unusual American Jones Live-Map Meter," http://www.christies.com/LotFinder/lot_details.aspx?fromsalesummary&intObjectID=719723.

43. "Jones Live Map" [advertisement], *Saturday Evening Post* 182.3 (26 February 1910): 44.

44. Touring Club of America, *Catalogue of Jones Live-Maps for the Jones Live-Map Meter* (New York: Jones Live-Map Meter Department, United Manufacturers, c. 1910).

45. *The Live-Map … What Happens Without It* (New York: Jones Live-Map Meter Co, 1910).

46. "Favors American Highway Congress," *New York Times*, 21 August 1910, C8.

47. According to the catalog, the club was the "compiler and exclusive distributor of Live-Maps"; see Touring Club of America, *Catalogue of Jones Live-Maps*, inside back cover. Distances, of course, were "measured by the Jones Speedometer-Odometer"; see *The Live-Map*, 29

48. "New Tour for Motorists to the Florida Peninsula," *New York Times*, 16 October 1910, 65.

49. Ernest Albert Jones, "Route Indicator Mechanism," US Patent 1,092,147, filed 29 July 1913, and issued 7 April 1914.

50. Examiner to Ernest Albert Jones (c/o Percy H. Moore [his attorney]). 13 September 1913. Signed, not typed, the examiner's name was too garbled to confirm either online or in the city directory.

51. Examiner to Ernest Albert Jones (c/o Percy H. Moore [his attorney]). 22 January 1914. Signed, not typed, the examiner's name was again too garbled to confirm either online or in the city directory.

52. See the European Patent Office website (espacenet.com). The South African patent was revealed when online searching discovered a legal notice in the *London Standard* for 13 October 1913, which listed his patent for "Improvements in or relating to Route Indicators."

53. Jones's parents had been born in Canada, moved to the United States around 1870, and lived in New York City, where his father, William Colpitts Jones, was a real estate agent. Because both brothers were born in June and July, respectively, it seems likely that the family summered in Saratoga, a popular resort. The family was not listed in the Saratoga city directories for 1876 and 1882. No intervening years were available through Ancestry.

54. As with his brother Joseph, I never found a complete obituary or biography for Ernest Albert Jones. Nonetheless, images of primary documents available through Ancestry Library Edition provide a few key biographical details. For example, on a passport application dated 27 May 1918, he stated that he was born in Saratoga, New York, on 1 July 1880, and was the husband of Frances Evelyn Johnston (a British subject) and father of Joseph Thornton Jones, born 28 June 1906. Because Ernest Albert was listed in the 1880 Census as a one-year-old, he was probably born in 1879, not 1880.

55. Ernest Albert Jones, "Improvements In or Relating to Kinematograph Target Apparatus," British Patent 1914-23307, priority date 30 November 1914, and issued 6 May 1915.

56. Ernest Albert Jones, "Improvements In and Relating to Driving Mechanism for Indicators or the like for Mechanically-propelled Road Vehicles," British Patent 1913-01825, priority date 22 January 1913, and issued 9 October 1913; and "Improvements in Carburetting Apparatus for Internal Combustion Engines," British Patent 1909-16215, priority date 12 July 1909, and issued 23 June 1910.

57. Johns-Manville discontinued the Jones name in 1916 and sold the assets to another speedometer manufacturer, Stewart-Warner, in 1924. Furstnow, *The Antique Automobile Speedometer*, 12, 15–17.

58. "The Jones Motrola," *The Popular Engineer* 12.1 (July 1919): 28.

59. Douglas, *Radio Manufacturers of the 1920s*, 77.
60. Joseph W. Jones, "Range Estimating Trainer," US Patent 2,364,720, filed 28 September 1943, and issued 12 December 1944. A patent for a similar device, filed a month earlier, was issued in August 1945.
61. Rhodes, "Route Indicator."
62. Perry, "Some Remarkable Mechanical Road Guides."
63. S. W. Mellotte (examiner) to Jay B. Rhodes (c/o Chappell & Earl [his attorneys]), 30 August 1910, and Jay B. Rhodes, "Route Indicator," US Patent 951,966, filed 6 July 1909, and issued 15 March 1910.
64. Bert O. Rhodes (assigned to Jay B. Rhodes), "Route Indicator," US Patent 1,010,802, filed 15 March 1909, and issued 5 December 1911.
65. According to diverse sources accessed through Ancestry Library Edition, Jay B. Rhodes was born 2 March 1865, and died on 12 October 1931, whereas his brother Bert was born on 25 July 1869, and died on 2 March 1943. According to 1880 Census records, Jay's original name might have been Byron J. Rhodes. Another source was "J. B. Rhodes: Kalamazoo's Edison," *Museography* [Kalamazoo Valley Museum] 6.2 (winter 2007): 14–15.
66. For example, see Kalamazoo Valley Museum, "Kalamazoo 'Patent King'—Jay B. Rhodes," http://www.kalamazoomuseum.org/calendar/display/index.php?event=181; also see Jay B. Rhodes, "Dispensing Can," US Patent 1,403,636, filed 27 December 1920, and issued 17 January 1922; and Jay B. Rhodes, "Safety Razor and Blade Therefor," US Patent 1,840,056, filed 7 April 1930, and issued 5 January 1932.
67. The device that recorded master disks was patented before the device that played published disks. Lee S. Chadwick, "Recording Means for Signaling Devices," US Patent 1,002,368, filed 17 March 1910, divided 18 April 1910, and issued 5 September 1911; and Lee S. Chadwick, "Signaling Device," US Patent 1,180,239, filed 17 March 1910, and issued 18 April 1916.
68. Sources differ on the type of material used for the disk. A 1953 retrospective says aluminum; see William D. Ellis, "Chadwick ...," *True's Automobile Yearbook* no. 2 (1953): 88–84, 132–33. By contrast, a contemporary source says paper; see "Development Briefs," *Motor Age* 21.11 (14 March 1912): 42–45, esp. 44.

69. For examples, see Arthur W. J. G. Ord-Hume, *Player Piano: the History of the Mechanical Piano and How to Repair It* (London: Allen & Unwin, 1970).
70. "Development Briefs," *Motor Age.*
71. J. G. Newton (examiner) to Lee Sherman Chadwick (c/o Howson & Howson [his attorneys]), 30 March 1910 (in the National Archives case file for Patent 1,180,239).
72. By contrast, the case file for the indicator patent, for which the number of claims was ultimately reduced from 17 to 9, has three letters of rejection and five letters with amendments.
73. The price of the Chadwick device is from an advertisement quoted (without attribution) by Ellis, p. 89.
74. "Development Briefs," *Motor Age.*
75. Chadwick died on 16 September 1958. For snippets of his life and accomplishments see Ken Bailey, "Hunter As Inventor," *Outdoor Canada* 33.5 (summer 2005): 99; Nick Baldwin and others, "Chadwick, USA 1903–1915," *The World Guide to Automobile Manufacturers* (New York: Facts on File, 1987), 89–90; Ellis, "Chadwick …"; Jeff Hartman, *Supercharging Performance Handbook* (Minneapolis, MN: 2011), 5–6; *Who's Who In Engineering* (New York: John W. Leonard Corporation, 1922), 249; and Stanley K. Yost, *The Great Old Cars: Where Are They Now?* (Mendota, Ill.: Wayside Press, 1960), 65–68.
76. Frank Feilhuber, "Road-Indicator for Vehicles," US Patent 951,637, filed 19 October 1908, and issued 8 March 1910. The patent lists Feilhuber as a resident of Litchfield, Connecticut, where he was apparently living when he filed. Several letters in the case file at the National Archives report a new place of residence, in Newark, New Jersey, to which his approved patent was mailed and which allowed me to correlate him with Census data available through Ancestry Library Edition. Born in Bavaria, Feilhuber was 46 years old when he applied for his patent. I could not find him in coding sheets for the 1940 Census, which has information about years of schooling.
77. Max Bremsy, "Mechanically Operated Road Map," US Patent 1,064,694, filed 8 September 1909, and issued 10 June 1913. About 50 years old when he applied for his patent, Bremsy died in 1923. He had legally changed his last name from Boremsky to Bremsy in 1906.

78. Max Bremsy, "Clothes-Line Clamp," US Patent 1,070,399, filed 30 July 1909, and issued 19 August 1913.

79. Permit 3591, according to the *Official Automobile Directory of the State of New York* (New York: J. R. Burton & Co., 1914), n.p.

80. George E. Boyden, "Chart for Vehicles," US Patent 1,113,747, filed 22 November 1911, and issued 13 October 1914. Boyden was about 44 years old when he applied for the patent.

81. George E. Boyden, "Vehicle Signaling System," US Patent 1,168,053, filed 17 July 1914, and issued 11 January 1916. Before the patent was issued, Boyden assigned rights to a trustee, Alexander P. Browne, a prominent Boston attorney who specialized in patents and copyrights. For a sketch of Browne's life, see "Alexander P. Browne, Noted Lawyer, Is Dead," *Boston Globe*, 8 July 1920, reproduced by FindAGrave.com.

82. George A. Deardorff, "Route Indicator," US Patent 1,050,434, filed 9 April 1912, and issued 14 January 1913.

83. George A. Deardorff, "Road Map for Automobiles," US Patent 1,068,112, filed 8 July 1912, and issued 22 July 1913.

84. George A. Deardorff, "Road Indicator," US Patent 1,250,811, filed 25 May 1916, and issued 18 December 1917.

85. George A. Deardorff, "Road Map," US Patent 1,184,726, filed 15 March 1915, and issued 30 May 1916.

86. Pio Papini, "Indicator for Illustrating and Signaling the Route of a Vehicle," US Patent 1,112,086, filed on 13 March 1911, and issued 29 September 1914.

87. Celora M. Stoddard, "Route-Indicator for Automobiles," US Patent 1,200,262, filed 16 February 1914, and issued 3 October 1916.

88. William E. Reilly, "Route Indicator," US Patent 1,1,75,992, filed 3 May 1915, and issued 21 March 1916. Reilly was born around 1893 and died on 29 December 1933.

89. Howard Cranmer, "Route Indicator," US Patent 1,246,240, filed 7 April 1916, and issued 13 November 1917. Cranmer was born in 1863 or 1864.

90. Henry A. Hubschmitt, Jr., "Route Indicator for Automobiles," US Patent 1,236,565, filed 19 October 1916, and issued 14 August 1917. According to his World War I draft registrations and other documents, Hubschmitt was born on 4 September 1885. According to Findagrave.com, he died in 1960.

91. Henry A. Hubschmitt, Jr., and Richard D. Spain, "School Furniture," US Patent 1,292,470, filed 19 October 1916, and issued 28 January 1919; and Henry A. Hubschmitt, Jr., "Roof Bracket," US Patent 1,562,965, filed 31 May 1924, and issued 24 November 1925.

92. William F. Brien and Marvin H. Whittaker, "Route and Station Indicating Means," US Patent 1,310,978, filed 21 January 1918, and issued 22 July 1919.

93. According to the image of his tombstone in a family plot in a Hoosick Falls, New York, cemetery (found on Findagrave.com), Whittaker was born in 1861 and died in 1951. According to his World War I draft registration, Brien was born on 31 March 1873 in Cherry Valley, New York. An August 1918 passport application reported his normal occupation as salesman but noted his plans to travel to Britain and France for "Y.M.C.A. war work." Manuscript records for the 1920 Census list him as "secretary" at the YMCA in Indianapolis.

94. Edward Siegel, "Vehicle Route Indicator Device," US Patent 1,350,244, filed 30 October 1919, and issued 17 August 1920. According to his World War I draft registration, Edward Pius Siegel was born on 9 December 1883 in Manhattan, and was employed by the Morse Dry Dock and Repair Company. According records from the 1940 Census, he had an eighth-grade education and ran his own plumbing business. Other public records confirm plumbing as his trade. At least one other inventor had the same first and last name, but the middle initial (P.) and place of residence (Elmhurst, Queens) mentioned in the patent point to this Edward Siegel as the inventor.

95. James R. Akerman, "Twentieth-Century American Road Maps and the Making of a National Motorized Space," in *Cartographies of Travel and Navigation*, ed. James R. Akerman, 151–206. (Chicago: University of Chicago Press, 2006).

96. Robert L. French, "Maps on Wheels: The Evolution of Automobile Navigation," in *Cartographies of Travel and Navigation*, ed. James R. Akerman, 260–90. (Chicago: University of Chicago Press, 2006).

CHAPTER 4

Folding, Unfolding

Although academic cartographers showed little interest in map folding, they didn't neglect it entirely. A brief examination of what they wrote provides a context for examining folding inventions registered at the US Patent Office.

Perhaps the first American academic cartographer to comment publicly on map folding is Erwin Raisz, who discussed it in a single short paragraph near the end of his classic 1938 textbook *General Cartography*. In a section titled "Preservation and Cataloguing of Maps," he included a simple three-part illustration captioned "The standard method of folding maps"—an assertion contradicted in the accompanying paragraph, which began "Maps are folded in various ways; there is no well-established convention for uniform folding."[1] In his example, three horizontal creases reduced a small, schematic map of the United States to a quarter of its size (Fig. 4.1). Three vertical creases then compressed the map to a sixteenth of its original size, with the map title appearing on top. These are called accordion or concertina folds because the map can be expanded or compressed like its namesake musical instrument. A quarter century later, in a substantial rewrite titled *Principles of Cartography*, Raisz reversed the sequence by starting with a set of five relatively long vertical creases, which better fit the accordion metaphor and made the map easier to hold than a horizontally elongated map. Reduced further by a pair of horizontal folds, the map also "fits a pocket better," he claimed.[2] No longer

© Mark Monmonier 2017
Mark Monmonier, *Patents and Cartographic Inventions*,
Palgrave Studies in the History of Science and Technology,
DOI 10.1007/978-3-319-51040-8_4

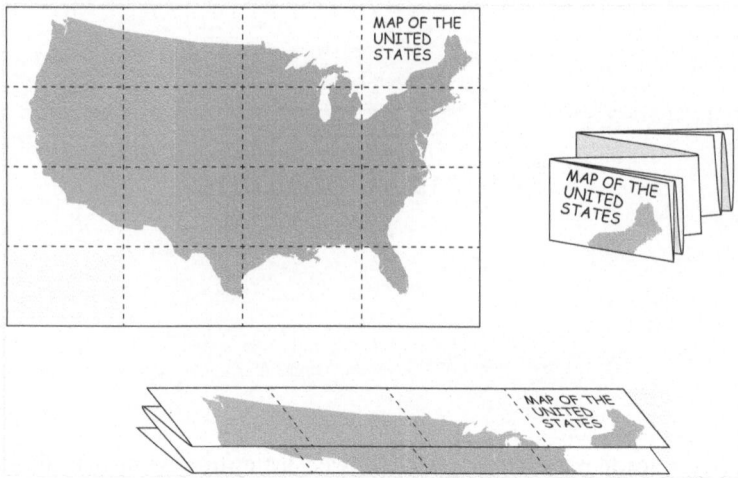

Fig. 4.1   Erwin Raisz's "standard" folding scheme used a succession of horizontal and vertical accordion folds to reduce a map to one-sixteenth its original size. Adapted from Raisz, *General Cartography* (1938), 345

ambivalent about what was common or standard, Raisz proclaimed that maps "should" be folded this way.

Meanwhile, another strategy was gaining prominence: a fold that involved cutting the paper so that several map sections could be viewed together without opening the full map. Raisz included one version of the cut fold in the second edition of his *General Cartography*, published in 1948, and the US Army described a similar strategy in its field manual on map reading, which emphasized the value of a small map that could be easily carried (Fig. 4.2).[3] Because Army maps were not slit at the printing plant, soldiers had to make their own cuts with a sharp knife or razor blade, preferably after practicing with scrap paper or an obsolete map.

Raisz's principal competitor in the textbook market, Arthur Robinson, had little or no interest in map folding, a topic not indexed (or otherwise mentioned, as far as I can tell) in any of the six editions of his *Elements of Cartography* published between 1953 and 1995, the last produced with four co-authors.[4] One of these collaborators, Phillip Muehrcke, had co-authored (and co-published) with his wife a 1978 textbook on map use that included a three-page section on map folding—a topic discarded by

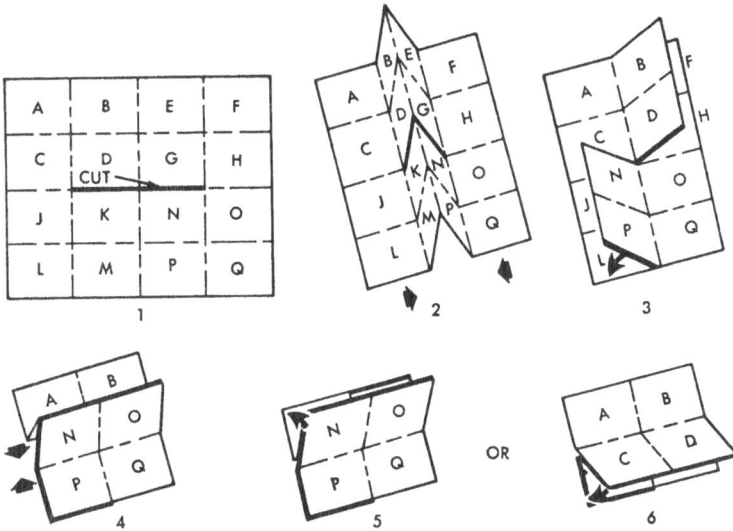

Fig. 4.2 Army Field Manual 21–26 showed how the soldier who partly split fold lines could view two or four adjoining sections without opening the full map. From Department of the Army, *Map Reading* (March 1965), 4

2012, when the seventh edition, prepared with two additional co-authors, appeared under another imprint.[5] Among the other cartographic texts oblivious to map-folding strategies is *Map Appreciation*, a 1988 release that I co-authored with George Schnell, a longtime collaborator—we mentioned folding once, but merely to contrast folded street maps and other hardcopy products with electronic maps.[6]

Despite a few gems, cartographic journals were similarly apathetic. Walter W. Ristow, who headed the Geography and Map Division at the US Library of Congress, mentioned folding toward the end of his classic 1964 *Surveying and Mapping* article on oil company road maps.[7] Although Ristow's observation that most of them employed an accordion fold seems accurate, his claim that "Almost one hundred different folds have been devised and patented" would not have been supported by the patents database in the early 1960s. In highlighting the persistent problem of refolding maps, he cited Scottish playwright James M. Barrie's delightful but obscure rant "Shutting a Map."[8] Better known as the author of Peter Pan, Barrie acknowledged a need for folded maps but denounced map publish-

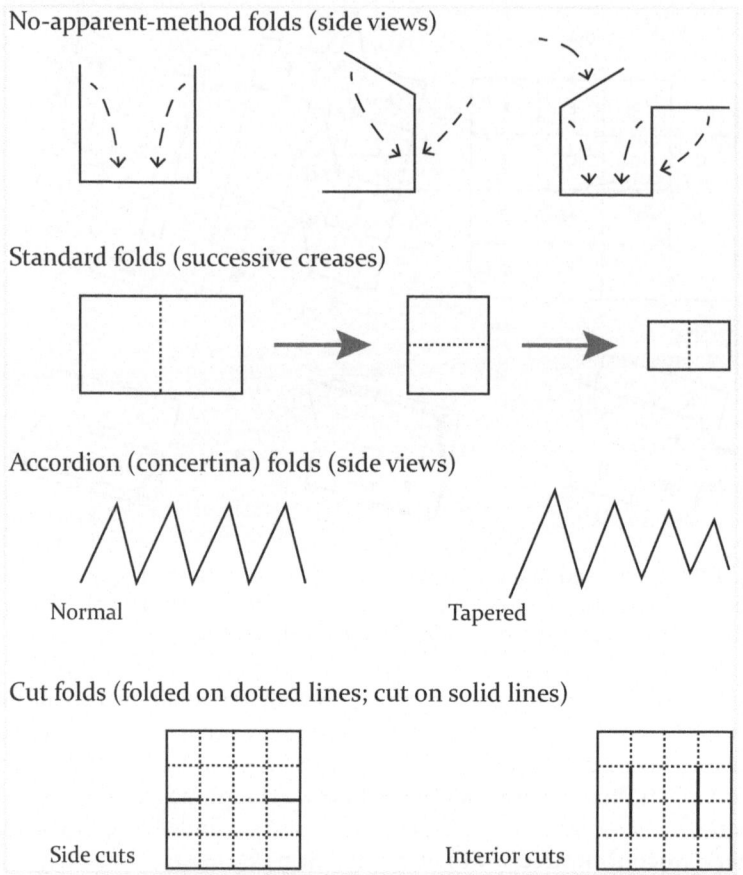

No-apparent-method folds (side views)

Standard folds (successive creases)

Accordion (concertina) folds (side views)

Normal   Tapered

Cut folds (folded on dotted lines; cut on solid lines)

Side cuts   Interior cuts

Fig. 4.3 Harold Otness recognized these four types of map folds. Adapted from drawings in Otness, "A Primer on Map Folds and Map Folding" (1974)

ers as merchants of frustration. "Prominent among the curses of civilization," he alleged, "is the map that folds up 'convenient for the pocket'."

Ristow's paper was one of two articles cited by Southern Oregon College library science professor Harold M. Otness, in his "Primer on Map Folds and Map Folding," a four-page paper published in 1974 in the Western Association of Map Libraries' *Information Bulletin*, one of cartography's more obscure journals.[9] Otness recognized four different kinds of folds, described schematically in Fig. 4.3. The first of these, the "no apparent method fold," he denounced as the "quickest and cheapest" accommodation of available folding machinery: an irrational strategy that produced an awkward map. His second method, labeled "standard," simply folded a map in half and in half again, and so on, with each new crease perpendicular to the preceding fold. If the folds were inward, an obvious pattern made it easy to refold the map, but the entire map had to be opened to view any part of it. By contrast, his third method, the accordion fold, made it possible to view part of the map without completely unfolding it. Although Otness's fourth method, the "cut fold," was similar to the approach described by Raisz and the Army field manual, he recognized the possibility of exterior cuts, which begin at the edge of the sheet, as well as interior cuts. Although a carefully chosen combination of cut folds and accordion folds could let a user view any small part of large map without opening it fully, cheap paper would crack after repeated opening and closing. An additional category, "miscellaneous folds," recognized other approaches inspired by origami, including folds similar to the dramatic three-dimensional "pop-up" illustrations in children's books.

The other article cited by Otness was "The Folding of Maps," by Helmut Mühle, who had apparently worked in printing or map publishing.[10] Published in Frankfurt in 1959 by the Institut für Angewandte Geodäsie (Institute for Geodesy), it was an awkward translation of a more complete German-language version that relied heavily on numerous drawings to explain the use, individually or in combination, of pleat and diagonal folds. Like other geometric endeavors, map folding is a topic not easily discussed with words alone—I know what a *pleat* is in a pair of trousers but need a diagram to understand the term's relevance to map folding.

Figure 4.4, which I adapted from drawings in Mühle's article, summarizes key strategies for reducing a map's size. The upper half shows a map divided into seven horizontal strips A–G (top left). Doubling over sections B and C forms one pleat and doubling over E and F forms another. The side view (top center) shows how these pleats substantially reduce the map's vertical extent. In this example, two sets of vertical cuts followed by an accordion fold then reduce the horizontal extent. Adhesive is applied to the back of sections A-1, D-1, and G-1 so that the map can be attached to a protective cover, convenient for an explanation and advertising. The lower half of Fig. 4.4 shows how diagonal folding with a single cut can reduce a square map to a triangle one-eight its original size.

In exploring varied strategies for folding maps, Mühle raised issues of design, cost, and durability. Although cut folds enhanced flexibility, map-

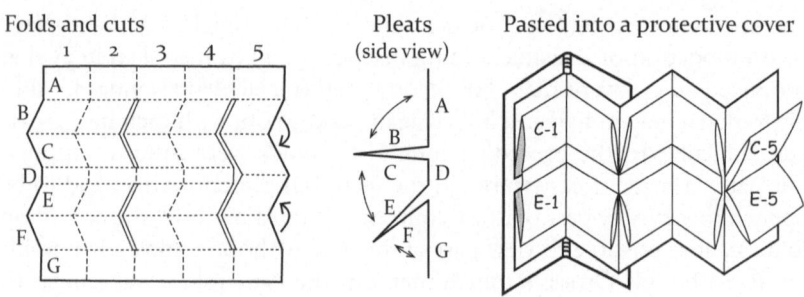

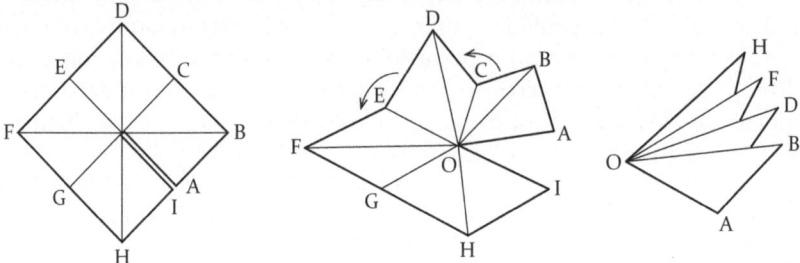

Fig. 4.4 Principles of pleat and diagonal folds, adapted from illustrations in Helmut Mühle, "The Folding of Maps" (1959)

makers had to make certain the cuts did not interrupt place and feature names. In addition, more durable (and thus more expensive) paper was required because the cuts increased the likelihood of tearing. Moreover, convenient folding made opening and refolding more frequent. Bending the paper in the first place was no less problematic: although bookbinding factories could cope with straightforward accordion folding, more complex folds were generally not suited to mass production.

Mühle was skeptical about patented folding schemes, which could increase manufacturing cost by as much as 30 percent and might not be practicable without a specially designed folding machine. In addition, copyright, trademark, and "the law of unfair competition" could be more useful to a map publisher than a patent, which typically ran for no more than 18 years. Noting that "up to now approximately 90 different foldings have been protected against plagiarism of any kind in Germany and abroad," he argued that "it will be very difficult to develop further appropriate foldings suitable for patent."[11] Little did he know.

British map dealer Lucinda Boyle underscored the importance of durability in her book *London: A Cartographic History, 1746–1950.*[12] Subtitled "200 Years of Folding Maps," it is a catalog of 498 folded maps collected by a recently deceased map collector. Almost all were cut into conveniently small rectangular sections and mounted on linen backing, which could be opened and refolded repeatedly, and generally lasted far longer than nonmounted maps. Although the sections were visibly separated along fold lines, flexible linen backing was especially useful when detailed travel maps had to be kept open to a particular section or pair of adjoining sections.

Boyle also wrote the entry "Folding Schemes" for the twentieth-century volume of the *History of Cartography*, in which she noted that mechanical folding machines were first used in the mid-1800s.[13] One of her two illustrations was a photograph of a clever 1995 star-shaped pop-out tourist map of Bath, England, and the other was the first page, with all of the drawings, of Gerhard Falk's 1951 US patent for the combination of cut and accordion folds that "became very popular" despite the "tendency of the slits along its folds to tear." She concluded by recognizing the zoomable electronic map as a twenty-first-century analog to the folded paper map.

One other academic author to specifically address map folding is Christopher Board, whose 1993 essay "Neglected Aspects of Map Design" also mentioned back-to-back map printing, protective covers, and the extension of large-scale topographic maps beyond the sheet lines to accommodate "important detail lying on the sheet boundary."[14] Board discussed the relative convenience of various folds used in government and commercial cartography, and praised the Miura-ori fold, an origami-based scheme devised by Japanese astrophysicist Koryo Miura to store solar panels for deployment in space, as "the most innovative folding system yet devised" and "probably the only example of applying theoretical research to map folding." Innovative, to be sure, but the only instance I've found of this technique in a cartographic patent is the "Wearable Folded Map," patented in 2003, which focuses on a reference guide strapped to the wrist, rather than a typical folded map.[15]

None of these authors looked systematically at map-folding patents. Any who did would surely have noticed the increased complexity of patents filed since the late 1980s, as shown in Fig. 4.5, a time-series graph for the number of pages or "drawing sheets" devoted wholly or partly to illustrations.[16] Each dot represents one patent, its horizontal position represents the year of filing, and its vertical position indicates the number

Number of drawing sheets, by year, for U.S. patents related to map folding

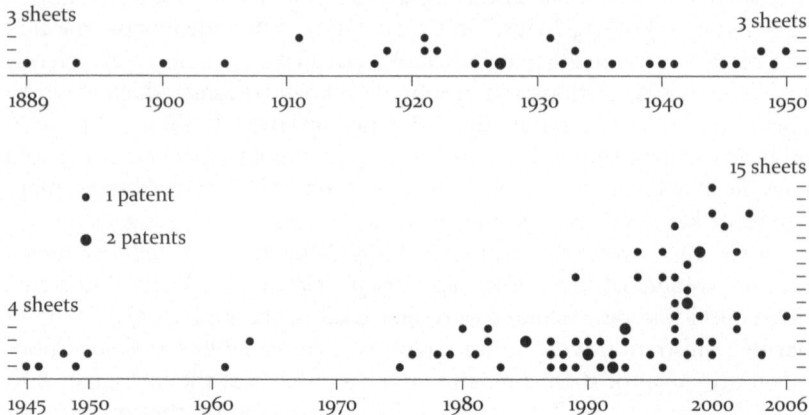

Fig. 4.5 Temporal trend in the number of drawing sheets for fold-related US patents classified as a Printed map (283/34) or an Indexed printed map (283/35). Each patent is represented by a dot positioned according to its year of filing and the number of pages devoted wholly or partly to illustrations. Compiled by author

of pages with illustrations. In 1921, for instance, two map-folding patents were filed, one with two drawing sheets and the other with three. The graph shows that until 1980 no map-folding patent had more than three drawing sheets. After 1990, however, the number of pages with illustrations was, on average, much higher. In general, the greater the number of drawing sheets in the patent, the more complex the folding scheme.

I confirmed this trend toward increased complexity with a second time-series graph (Fig. 4.6), for which I counted the number of drawings. Except for a patent with 22 drawings filed in 1933 (which took six years to process), the number of drawings remained comparatively moderate until 1994, when George Wallace McDonald filed a patent with 77 separately identified drawings and another with 30 drawings. A resident of the Channel Islands, McDonald filed seven successful patents for map-folding schemes, or slight modifications thereof, between 1989 and 1999. Because several of McDonald's patents and a few others filed between 1994 and 2003 would have been "off the chart," I condensed the height of the lower (more recent) half of Fig. 4.6 by adding two rows of dots: one representing 28 to 34 drawings and the other for 42 to 77 drawings. Too numerous to be dismissed as outliers, these extreme illustration counts

Number of drawings, by year, for U.S. patents related to map folding

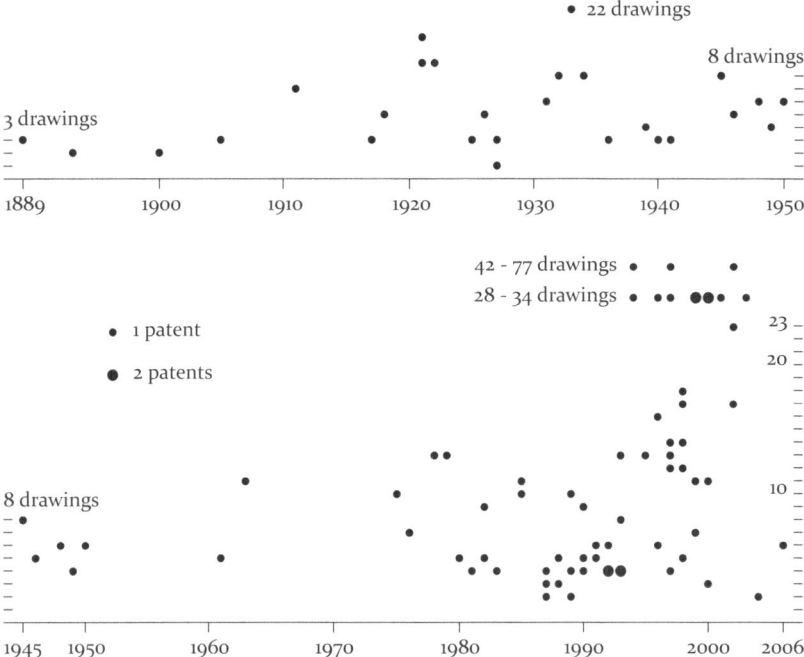

Fig. 4.6 Temporal trend in the number of separately numbered or lettered drawings in fold-related US patents classified as a Printed map (283/34) or an Indexed printed map (283/35). Each patent is represented by a dot positioned according to its year of filing and number of drawings. Compiled by author

reflect a larger trend, in the last decade of the twentieth century, toward an increased number of drawings.

Although patented map-folding schemes no doubt became more complicated over time because earlier inventors had already harvested the low-hanging fruit, their increased complexity no doubt reflects a shift toward more technically narrow and increasingly specialized drawings, starting in the 1980s, when the USPTO abandoned the long-held notion that a patent drawing should be legible at a glance to an intelligent member of the public. According to historian of technology William Rankin, this shift not only expanded the range of patentable inventions but also changed the notion of what constituted an invention.[17]

Another noteworthy development is the interest of inventors from other countries in map-folding schemes. As George McDonald's seven patents suggest, foreign inventors played a proportionately more prominent role in the increase in fold-related map patents after the mid-1980s. Of the 92-fold-related patents in my dataset, 33.7% (31) were awarded to foreign inventors, whereas only 12.7% (27) of the remaining 212 patents in the database went to non-US inventors. Figure 4.7, a time-series graph that distinguishes foreign from domestic inventors, shows non-US residents accounting for exactly half of the 26-fold-related patents filed between 1927 and 1979—significantly more than their one-third share of all fold-related patents—and 42% (19) of the 45-fold-related patents filed between 1987 and 2000. Although adjusting these time spans might dilute the percentages, it's clear that non-US inventors are proportionately overrepresented in the filing of successful map-related patents.

A geographic pattern is also apparent insofar as England and Germany, with 13- and 10-fold-related patents, respectively, were overrepresented in the non-US group, which includes patents filed by inventors in Japan, the Netherlands, and Sweden (2 each) as well as inventors in Canada and Switzerland (1 each). Although England and Germany are clearly more prominent, their counts reflect prolific inventors like George McDonald, who filed 7 of the 13 patents from Britain, and Gerhard Falk, who filed three of the ten patents from Germany.

Patents related to map folding, by year and domestic or international residence

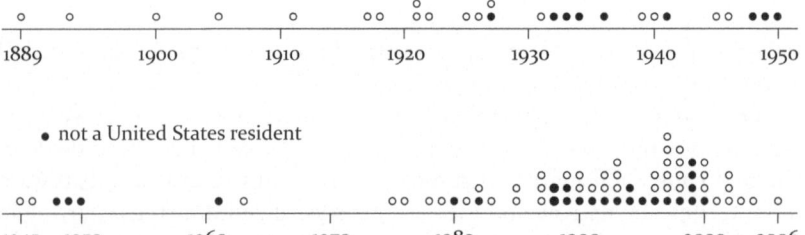

Fig. 4.7 Time-series graph comparing fold-related patents issued to foreign residents (*solid dots*) to those issued to US residents (*open dots*). Each dot, positioned horizontally according to year of filing, represents one US patent assigned to the Printed map (283/34) or Indexed printed map (283/35) category. Compiled by author

No less intriguing is the dearth of patents filed by foreign inventors between 1889 and 1931, when domestic inventors originated all but one of that period's 15 patents (Fig. 4.7). Although European inventors had been patenting map folds even earlier, they apparently had little interest in folded maps for the American market until 1927, when Friedrich Mattenklott, of Berlin, Germany, applied to patent a "Map Finder."[18] Indeed, folding was only a secondary feature of his invention, which consisted largely of two strips (coordinate axes) for indexing locations on a folded map attached to a notebook. One strip was attached to the map, and the other slid outward from the notebook, parallel to the binding. His patent had only one drawing.

As noted in Chap. 1, folding was not the dominant feature in many patents involving a folded map. Indeed, folding played only a subordinate role in the earliest patent in the map-folding dataset, for the "Educational Globe" invented by New York City resident Olin D. Gray.[19] Three figures on his single drawing sheet describe a small globe that opens to reveal a "strip of flexible material" containing pictures described on the globe by lines presumably relating to Columbus's voyages (Fig. 6.11). Although the strip nicely illustrates a simple accordion fold, the map is on the globe, not the strip. Gray filed his application in mid-November 1889, and because the invention was both straightforward and cleverly original, the Patent Office issued its approval expeditiously, on the last day of the year.

Markedly more sophisticated than Gray's invention is the "Book Fold Map" invented by Stacy E. Boyer, of Casper, Wyoming, and filed with the Patent Office in December 1922.[20] Boyer used nine drawings on two drawing sheets to describe the transformation of the hypothetical "Map of Doe County" at the top of his front page (Fig. 4.8) to the bound book-like folded map at the bottom of the second drawing sheet (Fig. 4.9). His initial map is partitioned into 48 oblong panels arranged in eight rows and six columns and numbered in accord with the meandering left-to-right-then-right-to-left sequence used in the US Public Land Survey. Dashed lines between the panels represent folds, bent to point upwards along the thicker lines and downward along the thinner ones; the side view in his second figure, below the map, describes this accordion fold. On the full map thick solid vertical lines from J to K and from L to M represent cuts in the paper. According to Boyer, "these slits permit the various folds of the map to be turned in the manner of pages when tracing routes, roads, and trails, etc., either laterally or longitudinally." It works—I tried it.

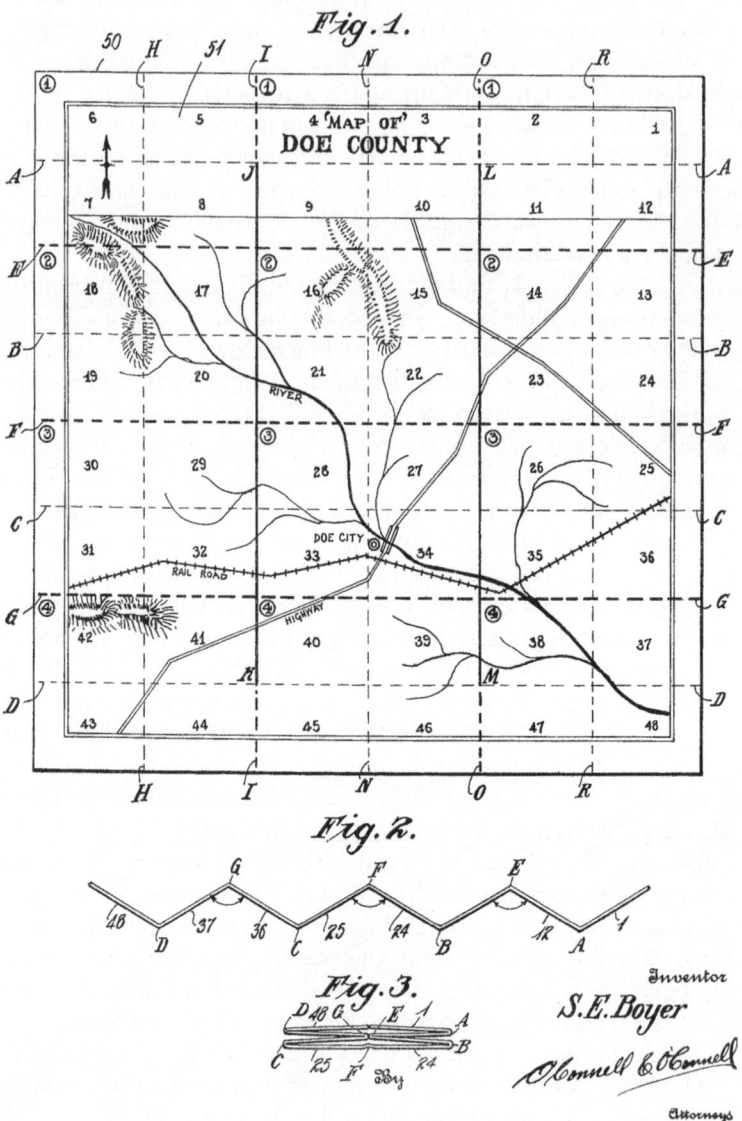

Fig. 4.8 Front page of Stacy Boyer's "Book Fold Map" includes a map sheet ("Fig. 1," at top) with 48 panels arranged in eight rows and six columns (US Patent 1,531,065; 1925)

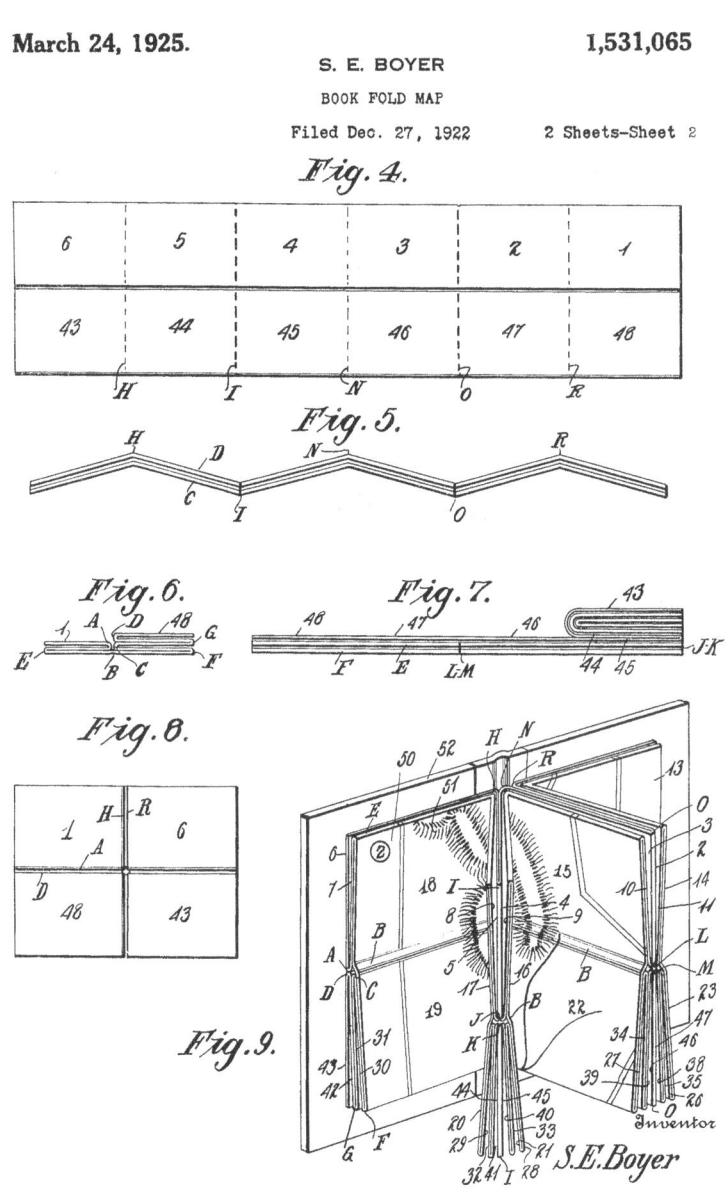

March 24, 1925.

S. E. BOYER

BOOK FOLD MAP

Filed Dec. 27, 1922

1,531,065

2 Sheets-Sheet 2

Fig. 4.

Fig. 5.

Fig. 6.

Fig. 7.

Fig. 8.

Fig. 9.

Inventor

S.E.Boyer

By

O'Connell & O'Connell

Attorneys

Fig. 4.9 Second drawing sheet for Stacy Boyer's "Book Fold Map" includes a folded map attached to a cover, like a book, open to panels 18 and 19 on the left and panels 15 and 22 on the right ("Fig. 9," at bottom right). Panel numbers correspond to the numbered cells in Boyer's first drawing (Fig. 4.8, top). Paste on the back of panels 1, 6, 43, and 48 holds the map to its cover (US Patent 1,531,065; 1925)

At the bottom right of the second drawing sheet (Fig. 4.9), an oblique view of the bound book map is opened slightly, to reveal two facing leaves, each with two of the original panels, elongated horizontally as on the initial map. Note panels 18 and 19 on the left and 15 and 22 on the right. Boyer described the map as anchored to the cover by adhesive on the backs of panels 1, 6, 43, and 48. This configuration of folds and leaves allows the user to move across the map by turning the leaves horizontally, like the pages in a book, or by flipping the lower pair of panels upward or the upper pair downward—of course the user must carefully avoid tearing the paper. Navigating to the desired panel requires patience, but there is no need to open the entire map and then face the frustration of refolding.

The specifications were less straightforward than the patent examiner would have liked. In a letter to Boyer's attorney, he claimed to have "constructed a map, in all respects, it is believed, in accordance with the directions given in the specifications."[21] But because the four adjacent panels at the upper right of the map could not be laid flat, he concluded that the "map fails to function," and declined to proceed with his evaluation until the inventor provided "a model of his map constructed according to the specification." I can sympathize: after struggling for over an hour with Boyer's words and images, I looked carefully at his initial and final drawings; made the specified cuts, upfolds, and downfolds; taped the backs of panels 1, 6, 43, and 48 to a card; and fiddled with the paper for the minute or so it took to make the scheme work. Reverse engineering can work wonders.

Boyer's attorneys (a Washington law firm) responded promptly with the requested specimen, which they asked the examiner to return when no longer needed—a pity because I would have been delighted to find a worked example in the case file at the National Archives.[22] A week later the Patent Office rejected 13 of the application's 16 claims, mostly because of similarity to existing British patents.[23] After roundly dismissing Boyer's outlandish claim to having invented "a foldable map having the back thereof coated with adhesive"—hardly innovative in the 1920s— the examiner allowed a wordy but concise statement that captured the essence of his invention, namely, "a map comprising a sheet provided with longitudinal and transverse fold-lines dividing the sheet into a plurality of quadrants, said sheet being folded in a zig-zag fashion on the said lines to bring the four corner quadrants together, said sheet being slit to permit the folded sections to be turned in the manner of leaves on both the longitudinal and transverse fold-lines." Concise, to be sure, but hardly a substitute for the accompanying drawings.

Probably pleased with the approval of what seems their key claim, Boyer and his attorneys responded a year later by renumbering the afore-mentioned allowed claim, slightly modifying two others, dropping nine of the rejected claims, adding four new claims, and defiantly ignoring four of the rejections.[24] I have no idea why they chose not to rewrite or drop the four claims that the examiner had deemed unacceptable. Whatever their strategy, it failed because when the Patent Office replied five months later, these four claims were "again rejected" on the basis of a 1911 US patent for a guidebook with multiple foldout map pages.[25] Even so, five of the remaining seven claims were "deemed allowable," and two others were judged acceptable if they agreed to some small but no doubt signifi-cant changes in wording. Satisfied with a narrower patent, Boyer and his attorneys promptly cancelled the four contested claims and accepted the examiner's recommended modifications.[26] After two additional correc-tions of "obvious informalities"—changes an examiner was empowered to make unless the applicant objected—Boyer's patent was approved, pend-ing payment of a final $20 fee.[27] The patent was issued in March 1925, 27 months after its filing.

Who was Stacy Boyer, and where did he get the idea for a clever book-like folding map? From federal records, city directories, and college year-books, among other sources, I know that he was born in 1901, attended the University of Colorado for three years, and was probably in college when he hired a patent attorney.[28] But I don't know what he majored in or why he left the university to return to Casper, Wyoming, to work as secretary-treasurer at his father's firm, the Western Blue Print Corporation. Although the company name suggests a focus on one-off engineering and architec-tural drawings, its capabilities were apparently more diverse and included printing large sheets that were folded, trimmed, and bound as books.

Later accomplishments confirm Boyer's business, entrepreneurial, and people skills. He married in his early 20s and around 1926 started his own business, S. E. Boyer & Co. (later the Prairie Publishing Co.), pos-sibly taking over from his father, who retired from printing to work as a salesman and later as city water commissioner. News articles indicate that he hobnobbed with editors and publishers throughout Wyoming and snagged state and school district printing contracts.[29] He enlisted in the Army in 1942, served in its aviation arm until 1946, and rose to lieutenant colonel in the Air Force Reserve. During the Eisenhower administration, he directed Federal Housing Agency activities in the state.[30] He died in 1992, and apparently never developed his invention.

Boyer did, however, create a map—a moderately large map, 27.5 × 16 inches, according to the *Catalog of Copyright Entries*—but its subject matter, "Salt Creek and Teapot Oil Fields, Natrona County, Wyoming," was hardly appropriate for sale to motorists or hikers in a compact format.[31] Published on 1 September 1922, less than four months before he filed his patent application, the map's inch-to-a-mile scale is consistent with his schematic example (Fig. 4.8, top). Although two copies were deposited with the Copyright Office, at the Library of Congress, I could not locate a specimen. What's relevant is that Boyer had not only engaged with mapmaking in his very early 20s but also engaged with the federal bureaucracy.

I also found a possible inspiration for Boyer's "Book Fold Map," but tight timing suggests it might be little more than a coincidence. A search of the Google Patents database using the keywords *fold, map,* and *slit* uncovered a patent titled "Book" issued to James G. Hall, of Burlingame, California, in mid-August 1922, less than five months before Boyer filed his patent application, and promptly published as a single-paragraph abstract and photoreduced drawing in the *Official Gazette of the United States Patent Office*, which might well have been available in the University of Colorado library.[32] What looks like only two rows of panels but many more than six columns seems a more rudimentary version of Boyer's "Book Fold Map," but the patent examiner apparently saw no interference between Boyer's expansive set of assertions and Hall's single claim, approved a mere seven months after filing. Hall saw his invention as useful for "a map or guide folded in book form and adapted to be carried in the pocket or for use by aviators when flying or motorists when touring without exposing a great area of the map to wind action."

As further evidence that the early 1920s was ripe for innovative map folds—in accord with the Theory of Multiples, discussed in Chap. 1—in June 1921 Robert S. Blair, of Sound Beach, Connecticut, near Stamford, asked to patent an invention that would provide "motor tourists and aviators" with a map "unaffected by exposure to strong winds and yet show the territory described thereon in comparatively large scale."[33] His "plurality of superimposed books" was essentially a thick book cover containing several folded maps or map booklets, each attached to a separate, comparatively stiff leaf; the leaves were hinged together to form a zigzag (accordion) fold. It is unlikely that Boyer or Hall would have heard of Blair's relatively complicated patent, which was not approved until August

1931, more than a decade after filing. I found no evidence that it was ever developed commercially.

Not all complicated folding schemes were the offspring of clever one-shot inventors who never developed their patents. The most prominent successful map-fold entrepreneur is Gerhard Falk, who was born in Berlin in 1922, and like Boyer had a flash of genius in his early 20s.[34] In 1938, after graduating from high school, Falk studied at the technical school for cartography run by the Reichsamt für Landesaufnahme (military survey) in Berlin. In 1939 he enrolled in the professional program for cartography at the city's graphic arts technical college, and after four semesters he became a certified cartographer. Drafted into the Wehrmacht in 1942, he served as a mapmaker on both the Eastern and Western fronts. According to folklore, his Eureka moment occurred in Hamburg in 1945, in the chaos of Germany's defeat, while either riding on a crowded streetcar or trying to find his way through a war-torn city with a cumbersome street map.

Falk's clever idea was a double-header: a map projection with a variable scale that accommodated more detail near the city center and a folded map that opened quickly to show just part of the city. He promptly set up a mapmaking company and in 1946 published his first map, for Hamburg. Falk Verlag became Europe's leading producer of city maps and survived its founder's death in 1978. As part of the publishing conglomerate MairDumont, the firm now offers an electronic route-planning service as well as paper maps.

Falk's two innovations called for different legal treatment. Because mathematical innovations are difficult to patent, his map projection was better suited to separate copyrights for individual maps. By contrast, his map-folding strategy was both innovative and patentable, as asserted (perhaps too boldly) by applications filed in Germany, Great Britain, and the United States. His latter filing, submitted in February 1949 through a New York attorney, was titled simply "Method of Folding Maps and the Like."[35] To reduce a large sheet of paper to a convenient size, he devised a series of formulas specifying the optimal number of concertina fold lines, cuts, and transverse pleats whereby a map reduced to book-like form could be opened outward, as shown in the last of four figures on his patent's front page (Fig. 4.10, right). The user should be able to unfold the map "quickly and without trouble," as the outstretched hands imply, and refold it with equal dispatch. In addition, "the section of the map being read should be as large as possible."

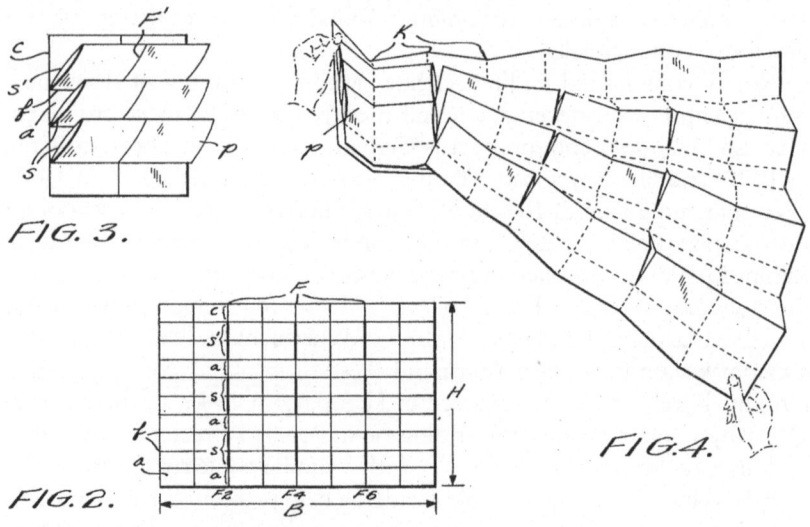

Fig. 4.10 Three of the four drawings for Gerhard Falk's first US patent. Heavier lines in his pre-fold layout (*bottom left*) represent slits (US Patent 2,572,460; 1951)

Falk's application did not fare well at the Patent Office. Because of language difficulties, all of his six claims were rejected as "indefinite and confused," and all but the fifth were "rejected as claiming subject matter not within the provisions of the patent statute"—the computational steps, however novel in their use of algebra, did not "perform a physical change or effect thereby transforming the subject matter to a different physical state."[36] Finally, after two years of amendments, the patent was provisionally approved, with only three, substantially reworked claims and two further amendments by the examiner.[37]

As noted earlier, Falk filed three of the ten map-fold US patents awarded to German citizens. His second application, submitted in March 1950 (while the first was still pending), involved claims and drawings more characteristic of what cartographers consider the Falk technique.[38] Titled "Map Capable of Being Folded Together and Spread Flat Again," it was submitted through a Detroit law firm that confessed to working "on instructions from a German patent associate."[39] Even so, all six initial claims were rejected "as drawn to matter insufficiently disclosed," and the patent examiner insisted, "The specification must be revised to conform

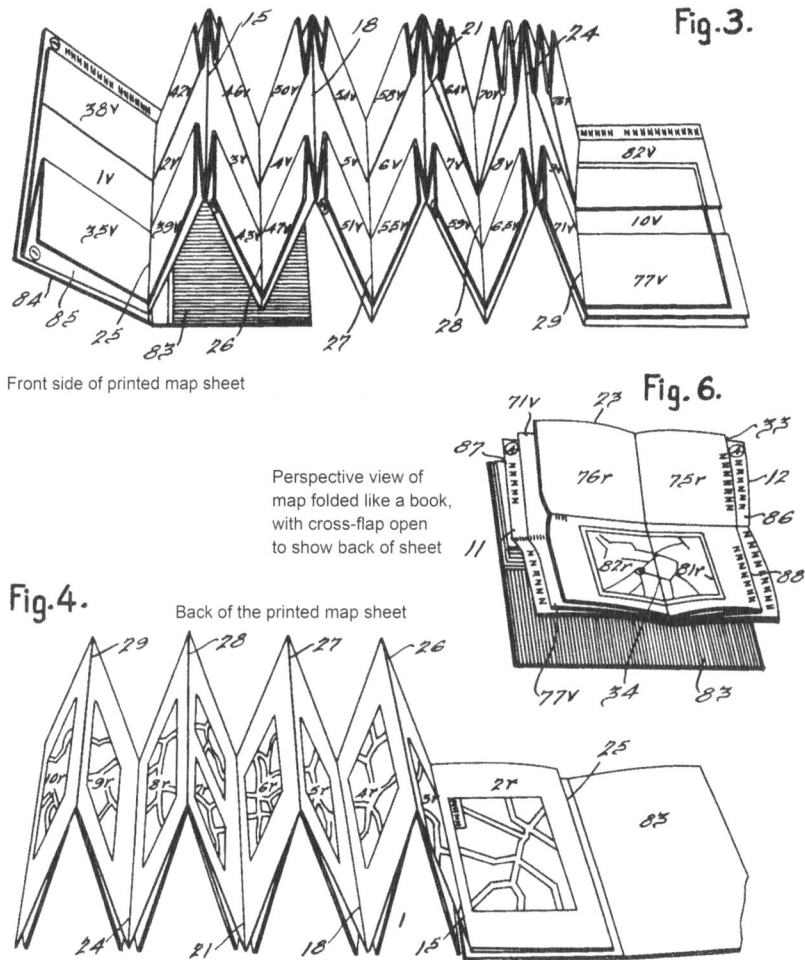

Fig.3.

Front side of printed map sheet

Fig. 6.

Perspective view of
map folded like a book,
with cross-flap open
to show back of sheet

Fig.4.

Back of the printed map sheet

Fig. 4.11 Three of the six drawings for Gerhard Falk's second US patent (US Patent 2,572,460; 1951)

with idiomatic English usuage [sic]." But two and a half years after filing and multiple amendments, Falk's application was approved, albeit with only two, suitably revised claims.

Falk's drawings (Fig. 4.11) describe a sheet printed on both sides, with a large map on the front and small detailed maps of neighborhoods or small cities on the back. Panels on the back can also include descriptive text and photos. A perspective view (middle right in Fig. 4.11) shows a book-like fold opened to a detail map on the back. Index information in the outer margins identifies the small maps and helps users locate specific places.

In July 1961 Falk filed a third patent, titled "Method for Folding Large Area Maps into the Shape of a Book and Correspondingly Folded Maps."[40] A modification was needed because the slits used in his earlier folding technique "cannot be accomplished mechanically … which raises the manufacturing costs considerably." His new folding technique accommodated a folding machine with "perforating knives," a "counter-pressure roller," "two blades arranged in parallel relation," and a strategy whereby "trapped air may escape through the perforations." At the bottom of this third patent, the examiner added a reference to the "Book Fold Map" patented in 1925 by Stacy Boyer, who probably never dreamed of a suitably sophisticated folding machine.

Falk's hope of an automated assembly line was premature. Until the mid-1980s, when Falk Verlag patented a folding machine invented in house by Alfred Vogtländer, the maps still had to be folded manually, "sheet by sheet," because the transverse cuts required for the prototypical Falk fold had made the large sheets "unstable [and] difficult to handle."[41] But this time a patented apparatus was able to reproduce the desired fold.

In contrast to Falk, who devised a comparatively complex technique for making a large map compact, British inventor George Wallace McDonald found a simpler way to make a somewhat smaller sheet fit a shirt pocket. He called his invention the Z-Card: Z for the second of its two concertina folds (also known as Z-folds) and *Card* for credit card—an apt metaphor even though a Z-Card map is closer in size to a wallet than to the credit cards within. Straightforward drawings (Fig. 4.12) in his patent show how five vertical folds followed by two horizontal folds can shrink a map by 94 percent. Moderately stiff front and back covers pasted to the backs of the upper-left and lower-right panels (items 16 and 18 in drawings 1, 3, and 4) not only advertise and protect the map but also serve as handles for opening and closing the sheet. Relatively few folds and an absence of slits make a Z-Card map durable, intuitive, and easy to manufacture.

McDonald, who was born in Scotland in 1955, got the idea in his mid-30s, while consciously trying to think up a product that was innovative, useful, and more profitable than writing travel guides, which he had been doing successfully since the early 1980s.[42] His business strategy

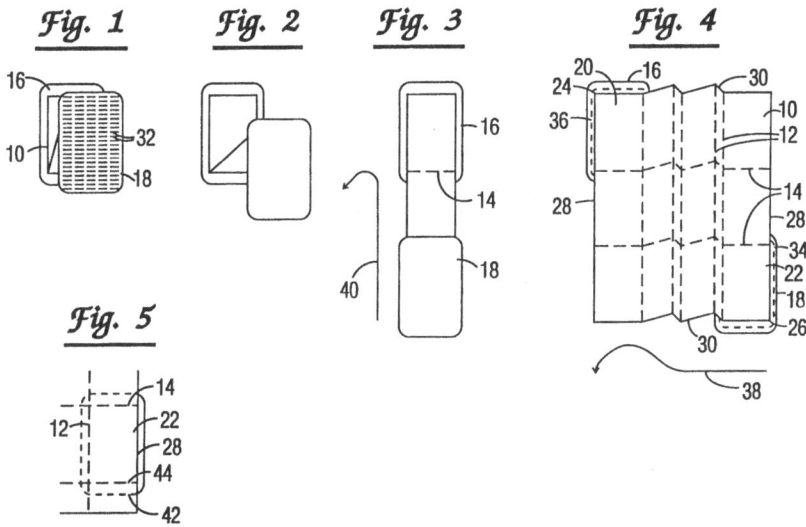

Fig. 4.12 Five straightforward drawings in George McDonald's patent concisely describe the two sets of accordion folds of a map with 18 panels (two attached to stiff front and back covers) that fits in a pocket and conveniently unfolds and refolds (US Patent 5,156,898; 1992)

involved securing patents in Europe and the United States and licensing them to franchisees in diverse markets. He began filing patents in 1986, and received his first US patent in October 1992, 16 months after filing. Its 16 claims mention a diverse range of modifications and uses, including dictionaries, phrases for travelers, advice for taking pictures and watching birds and trees, and first-aid instructions as well as maps.[43] Eager to expedite production and control costs, McDonald filed a patent in 1994 for a "Method and Apparatus for Providing Folded Sheets with Stiffeners."[44] Although the application took five years to process, its 22 claims were no doubt useful in warding off competition. Low-cost manufacturing is important because most Z-Cards, unlike Falk maps, are mass-produced for free distribution by tourist bureaus, attractions, and other advertisers.

One final noteworthy invention is the starburst map, which pops out when its wallet-like cover (or backing sheet) is opened, lies more or less flat when the cover is fully opened, and folds back when the cover is closed. As the name implies, the map bursts upward and radiates toward the rectangular sheet's four corners. A clever device with roots somewhere in the unwritten history of origami, it's enormously impressive the first time you see it. Unlike most exercises in origami, the opened-up state is not intended to be three-dimensional, but pressing the paper flat will weaken it along the creases. If crafted properly, the quasi-open map is highly readable.

No less fascinating is the numerous times the starburst map fold had been patented, with each inventor tweaking the process and crafting claims sufficiently distinctive to avoid direct conflict with a prior patent. The earliest complete patent I found was filed in February 1957 by Albert Soffa and Frederick Kulicke, Jr., independent inventors working under contract to Walter S. Sachs & Co., a Philadelphia investment firm specializing in fossil fuels. Twenty-one drawings on the eight drawing sheets for their "Automatic Folding and Binding Machine" describe an apparatus for "automatically impressing pleats or folds into a scored flat sheet [and applying a] backer whereby the bound sheet may thereafter be completely unfolded simply by opening the backer, or collapsed into a fraction of the unfolded state merely by closing the backer."[45] Their last drawing sheet (Fig. 4.13) shows the printed sheet's intended creases (*upper left*) and its not-quite-flat unfolded state (*middle left*).

Aside from a likely profit motive, I have no idea why Walter Sachs backed the two inventors or whether he ever manufactured a folding machine or licensed the patent.[46] A decade younger than his collaborators, he had no apparent experience in advertising, design, mapmaking, printing, or publishing, or any obvious interest in manufacturing, aside from the oil and gas industry. Soffa and Kulicke had been co-workers at a Philadelphia engineering firm in 1951, when they started their own company in Soffa's garage, and Kulicke & Soffa Industries later became prominent in the semiconductor industry. Kulicke, the firm's CEO, was born in 1917 and had been a company commander in the Normandy invasion during World War II.[47] His obituary did not mention college training, but coding sheets for the 1940 Census report four years of high school and employment as a draftsman. By contrast, Soffa, who was born in 1920, held a master's degree in mechanical engineering from Harvard.[48] The only clear connection is that they all lived in Philadelphia.

Although their patent never uses the word *map*, it identifies a clear precedent: the patent for a "Foldable Sheet" awarded in 1950 to Anders Oswald Palm, of Stockholm, Sweden.[49] Palm unabashedly asserted in his first sentence that his "invention relates to foldable sheets to be used as foldable maps, diagrams, programs, etc.," and his examiner cited Boyer's

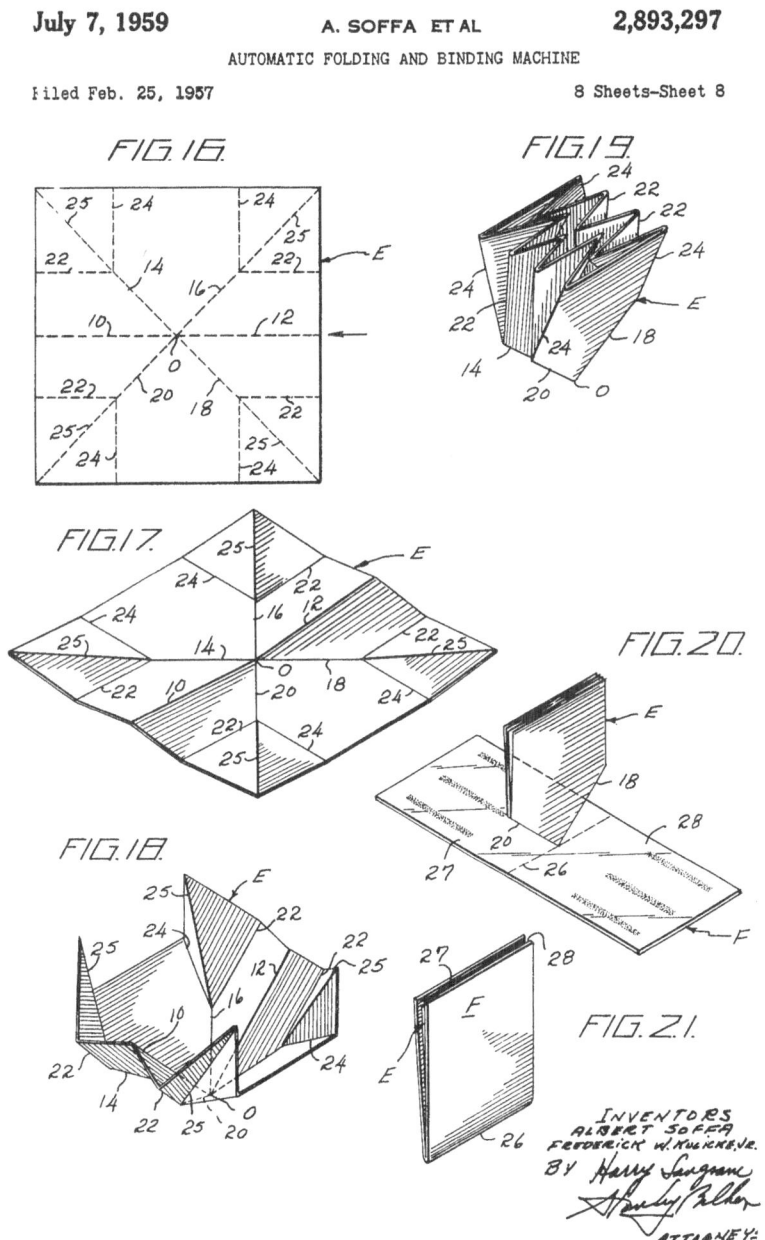

July 7, 1959

A. SOFFA ET AL

2,893,297

AUTOMATIC FOLDING AND BINDING MACHINE

Filed Feb. 25, 1957

8 Sheets–Sheet 8

Fig. 4.13 Drawing sheet 8 for Albert Soffà and Frederick W. Kulicke, Jr.'s patent for a folding machine able to produce a starburst fold (US Patent 2,893,297; 1959)

1925 "Book Fold Map" as the sole reference—another instance of the patent system as a distinctive literature. According to Soffa and Kulicke, Palm's invention, however clever and useful, was lacking because "the folds or pleats had to be formed or impressed by hand, thereby requiring a time consuming expensive operation." Their improvement was not the fold per se but a machine that reproduced it efficiently.

Figure 4.14 underscores the versatility and persistence of the starburst fold as a patentable cartographic invention. Note the similarity in fold lines between key drawings in patents issued to Palm in 1950, Soffa and Kulicke in 1959, Irving Sheroff and Howard Berwanger in 1973, Alfred and Elsbeth Vogtländer in 1987, Stephan Muth and Guenter Vollath in 1989 and 1990, and Derek Dacey in 2013.[50] Despite significant differences in the wording of their claims—and a few small differences in the drawings—all provide a map that unfolds and refolds quickly and usefully. All of them work, and all met the originality threshold at the Patent Office.

Like Palm and the Sachs collaborators, New York inventors Sheroff and Berwanger saw their invention languish. By contrast, the Vogtländers, who lived in Waldbröl-Hermesdorf, in the Federal Republic of Germany, assigned their patent to Falk Verlag fur Landkarten und Stadplane Gerhard Falk GmbH, and British inventor Dacey assigned his patent to Compass Maps Ltd, of Bristol, UK, which he had founded in 1993. His firm, which held other patents, mostly in Britain, and is now owned by an Atlanta, Georgia, corporation, offers PopOut Maps for over 50 cities in Europe, Asia, and the United States.[51]

In a similar move, New York inventors Muth and Vollath assigned their patents to mapmaker VanDam, Inc. Turns out that Muth is actually Stephan Van Dam, the firm's German-born owner and a graduate of the Parsons School of Design.[52] Inspired by a fellow student's folding metal sculptures, he conceived "Unfolds Maps"—a registered trademark—and built a business around it with the help of venture capitalists. In the early 1980s, he filed a patent on his own for the folding technique but chose to collaborate with engineer Vollath in devising a folding machine.[53] "The fold started it all," he told me, and even though the patents have expired, "30 years later we still use the same setup." In addition to Unfolds Maps, Van Dam makes StreetSmart Maps, which employ a simple accordion fold, and Pop-Up Maps, a starburst design that competes with Dacey's PopOut Maps in several markets. Although Pop-Up and PopOut maps both rely on a starburst fold, their designs and content are distinctive.

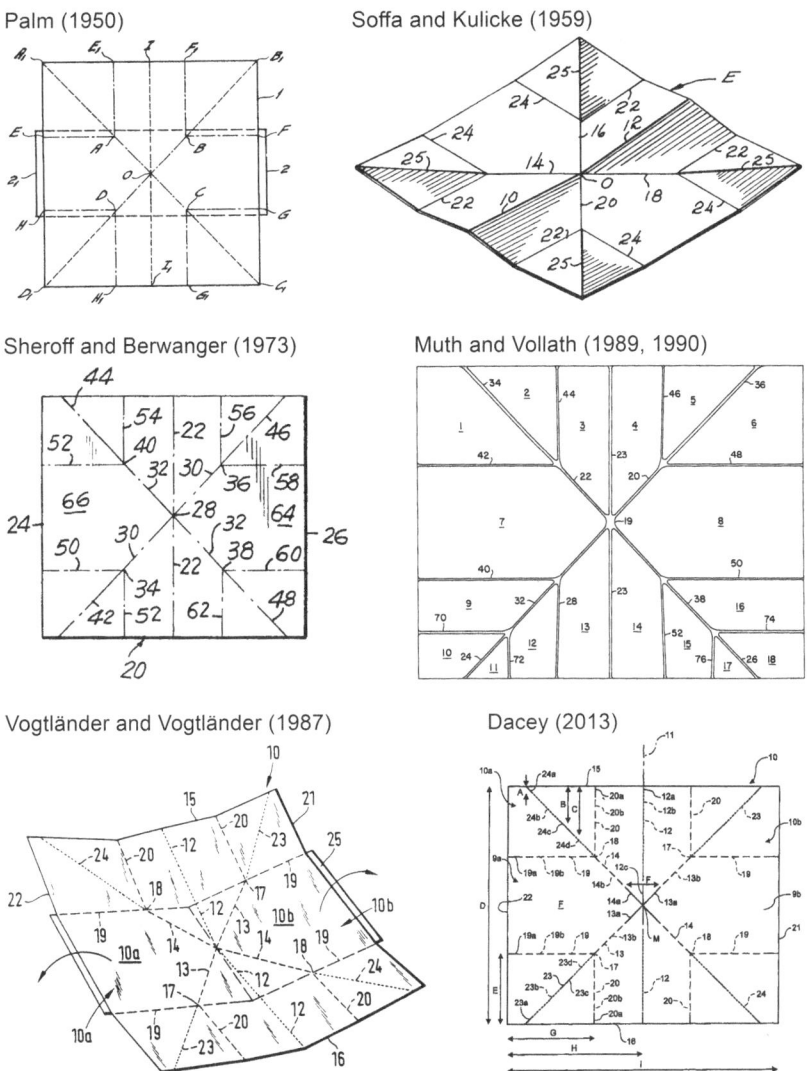

Fig. 4.14 Drawings with similar starburst fold lines for patents issued between 1950 and 2013 (*Upper left*: Fig. 1 from US Patent 2,525,937; 1950. *Upper right*: Fig. 17 from US Patent 2,893,297; 1959. *Middle left*: Fig. 15 from US Patent 3,753,558; 1973. *Middle right*: Fig. 1 from US Patents 4,826,212 and 4,917,405; 1989 and 1990. *Lower left*: Fig. 7 from US Patent 4,636,192; 1987. *Lower right*: Fig. 5 from US Patent 8,414,300; 2013)

Map folding entered the twenty-first century as a two-tiered pursuit: while the conventional accordion fold soldiered on as an accepted format for traditional, large-sheet folded street and road maps, the newer Z-Card and starburst folds became a convenient vehicle for smaller, pocket-sized maps focused on a city's built-up area or narrow themes like shopping or landmark buildings. All of these map-folding strategies came to rely upon innovative machines customized for efficient mass production: a parallel stream of inventing and patenting that bolstered the wave of map-fold patenting after 1985 (Fig. 1.6). With convenient compactness achieved, the wave subsided and innovative map folding became a dying art as patents expired and clever entrepreneurs turned to copyright's more enduring protections.

NOTES

1. Erwin Raisz, *General Cartography* (New York: McGraw-Hill, 1938), 345.
2. Erwin Raisz, *Principles of Cartography* (New York: McGraw-Hill, 1962), 117.
3. Erwin Raisz, *General Cartography*, 2nd ed. (New York: McGraw-Hill, 1945), 323–24; and Headquarters, Department of the Army, *Map Reading, Field Manual no. 21–26* (Washington, DC: 1965), 3–5.
4. Arthur H. Robinson, *Elements of Cartography* (New York: John Wiley and Sons, 1953); and Arthur H. Robinson, Joel L. Morrison, Phillip C. Muehrcke, A. Jon Kimerling, and Stephen C. Guptill, *Elements of Cartography*, 6th ed. (New York: John Wiley and Sons, 1995).
5. Phillip C. Muehrcke and Juliana O Muehrcke, *Map Use: Reading, Analysis, Interpretation* (Madison, WI: JP Publications, 1978); and A. Jon Kimerling, Aileen R. Buckley, Phillip C. Muehrcke, and Juliana O Muehrcke, *Map Use: Reading, Analysis, Interpretation*, 7th ed. (Redlands, CA: Esri Press, 2012).
6. Mark Monmonier and George A. Schnell, *Map Appreciation* (Englewood Cliffs, NJ: Prentice-Hall, 1988), 395.
7. Walter W. Ristow, "A Half Century of Oil-Company Road Maps," *Surveying and Mapping* 24 (1964): 617–37, esp. 636–37.

8. James M. Barrie, "Shutting a Map," in his *An Auld Licht Manse and Other Sketches* (New York: John Knox and Company, 1893), 113–23; quotation on 113.

9. Harold M. Otness, "A Primer on Map Folds and Map Folding," *Information Bulletin, Western Association of Map Libraries* 6.1 (1974): 13–16.

10. Helmut Mühle, "The Folding of Maps," *Informations Relative to Cartography and Geodesy*, Series II: German Contributions in Foreign Languages (Frankfurt am Main: Institut für Angewandte Geodäsie, 1959), 55–69.

11. Mühle, "The Folding of Maps," 56.

12. Lucinda Boyle, *London: A Cartographic History, 1746–1950: 200 Years of Folding Maps* (Wycombe, Buckinghamshire: Countrywide Editions, 2002).

13. Lucinda Boyle, "Reproduction of Maps: Folding Strategies," in *Cartography in the Twentieth Century* (Vol. 6 of *The History of Cartography*), ed. Mark Monmonier, 1336–38 (Chicago: University of Chicago Press, 2015); quotation on 1337.

14. C. Board, "Neglected Aspects of Map Design," *Cartographic Journal* 30 (1993): 119–22; quotations on 119 and 121, respectively.

15. Phillip Seawright, "Wearable Folded Map," US Patent 6,644,694, filed 1 July 2002, and issued 11 November 2003.

16. Before around 1970, when the Patent Office phased in a standard front page that included classification information and other data about the inventor and the patent, all of the illustrations were at the front of the patent on separate drawing sheets. In addition to these bibliographic data, the new front page included a single drawing deemed to reflect the entire patent. Sometimes much reduced in size to fit the space available, this key drawing was repeated later, in its appropriate position in the drawing sheets, all of which preceded the specifications and claims.

17. William J. Rankin, "The 'Person Skilled in the Art' Is Really Quite Conventional: U.S. Patent Drawings and the Persona of the Inventor, 1870–2005," in *Making and Unmaking Intellectual Property: Creative Production in Legal and Cultural Perspective*, ed. Mario Biagioli, Peter Jaszi, and Martha Woodmansee, 55–75 (Chicago: University of Chicago Press, 2011); esp. 57, 69–72.

18. Friedrich Mattenklott, "Map Finder," US Patent 1,755,742, filed 14 June 1927, and issued 22 April 1930.
19. Olin D. Gray, "Educational Globe," US Patent 418,455, filed 18 November 1889, and issued 31 December 1889.
20. Stacy E. Boyer, "Book Fold Map," US Patent 1,531,065, filed 27 December 1922, and issued 24 March 1925.
21. C. L. Wolcott (examiner) to O'Connell & O'Connell, 17 May 1923.
22. O'Connell & O'Connell to Commissioner of Patents, 21 May 1923.
23. J. F. MacNab (examiner) to O'Connell & O'Connell, 1 June 1923.
24. Dated exactly one year minus one day after the letter of rejection had been mailed, their response was received at the Patent Office the same day. Frank O'Connell to Commissioner of Patents, 31 May 1924.
25. M. K. Peck (examiner) to O'Connell & O'Connell, 3 November 1924. The patent cited in rejecting once more the four unacceptable claims was Herbert H. Patton, "Multipartite Multifold Map," US Patent 1,009,687, filed 17 August 1911, and issued 21 November 1911.
26. Frank O'Connell to Commissioner of Patents, 5 November 1924.
27. Thomas E. Robertson (Commissioner of Patents) to Stacy E. Boyer, 15 November 1924. Notifications of "obvious informalities" corrected by examiner M. K. Peck were dated 15 November 1924 (the same day) and 12 March 1925, less than two weeks before the patent was officially issued, on 24 March 1925.
28. My key sources, mostly accessed through Ancestry Library Edition, were enumerators' schedules for the Census of Population for 1910, 1920, 1930, and 1940; R. L. Polk city directories for Casper, Wyoming, available for 1924, 1925, 1928, 1929, 1934, 1937, 1939, 1941, 1943, 1945, and 1960; and the *Coloradan* (University of Colorado yearbook) for 1920 (p. 207) and 1921 (p. 224). Boyer was a member of the Class of 1923, but reported only three years of college to the 1940 Census.
29. For example, "Wyoming State News," *Star Valley Independent* [Afton, Wyoming], 30 March 1933, n.p.; "Editors Gather at Casper, Wyo.," *Billings Gazette*, 25 January 1936, 10; and "Wyoming FHA Director Named," *Billings Gazette*, 30 September 1953, 18; all in NewspaperArchive.com.

30. In addition to the aforementioned news articles, see *Official Register of the United States: Persons Occupying Administrative and Supervisory Positions in the Legislative, Executive, and Judicial Branches of the Federal Government, and in the District of Columbia Government* for 1954 through 1959.

31. Library of Congress, Copyright Office, *Catalog of Copyright Entries*, Part 1: Books, Group 2, n.s., vol. 19, no. 11 (Washington, DC: Government Printing Office, 1917), 1647.

32. James G. Hall, "Book," US Patent 1,426,291, filed 16 January 1922, and issued 15 August 1922; and James G. Hall, "1,426,291—Book," *Official Gazette of the United States Patent Office* 301 (15 August 1922): 557.

33. Robert S. Blair, "Map Construction," US Patent 1,820,115, filed 8 June 1921, and issued 25 August 1931.

34. Biographical sources for Gerhard Ernst Albrecht Falk include Hans Ermel, "Gerhard Falk," *Kartographische Nachrichten* 29 (1979): 32–33; "Falk," MairDumont [English-language], accessed 25 July 2015, http://www.mairdumont.com/en/marken-produkte/brands/falk/; "Falk (Verlag)," Wikipedia, last modified 3 May 2015, https://de.wikipedia.org/wiki/Falk_(Verlag); "Gerhard Falk," Wikipedia, last modified 18 July 2015, https://de.wikipedia.org/wiki/Gerhard_Falk; and Claudia Loebbecke and Claudio Huyskens, "Online Delivery Context: Concept and Business Potential," in *E-Commerce and V-Business: Digital Enterprise in the Twenty-First Century*, 2nd ed., ed. Stuart Barnes, 23–42, esp. 34–39 (Amsterdam: Elsevier, 2007).

35. Gerhard Ernst Albrecht Falk, "Method of Folding Maps and the Like," US Patent 2,572,460, filed 10 February 1949, and issued 23 October 1951. Chapter 5 discusses Falk's map projection patent, "Photographic Method for Making Geographic Maps," US Patent 2,650,517, filed 10 February 1949, and issued 1 September 1953.

36. M. V. Brindisi (examiner) to Michael S. Striker, 2 November 1949.

37. Brindisi to Striker, 11 April 1951.

38. Gerhard Ernst Albrecht Falk, "Map Capable of Being Folded and Spread Flat Again," US Patent 2,615,732, filed 14 March 1950, and issued 28 October 1952.

39. Neal A. Waldrop, Jr. (of Harness, Dickey & Pierce) to Commissioner of Patents, 29 December 1950.

40. Gerhard Falk, "Method for Folding Large Area Maps into the Shape of a Book and Correspondingly Folded Maps," US Patent 3,143,363, filed 11 July 1961, and issued 4 August 1964.

41. Alfred Vogtländer, "Process and Apparatus for Folding a Sheet Longitudinally and Transversely," US Patent 4,571,237, filed 1 June 1984, and issued 18 February 1986. At the time of the grant, the patent was assigned to Falk Verlag für Landkarten und Stadtplane Gerhard Falk GmbH, Hamburg, Fed. Rep. of Germany. The invention was also registered as German patent 3,320,731.

42. Biographical sources for George Wallace McDonald include Nick Clayton, "Z-Cards Unfolds a Fortune," *The Scotsman*, 30 July 1997, 7; "George Wallace McDonald," Wikibin, accessed 7 July 2015, John Frank-Keyes, "Pocket-Sized Card Proves Dream Idea," *South China Morning Post* (Hong Kong), 26 January 1994, supplement, 8; Liz Love, "The Z-Card—An International Success Story," *Business Works* [www.biz-works.net], 26 January 2012; and "Z-CARD," Wikipedia, last modified 1 March 2015, https://en.wikipedia.org/wiki/Z-CARD. Wallace left his "day job" as a travel writer after producing "a series of Z-CARDs for British Airways Club Europe." In 1992 he set up an international franchise firm "based around Z-CARD and its associated patented machinery," George McDonald, email correspondence, 31 July 2015.

43. George W. McDonald, "Folded Sheet," US Patent 5,156,898, filed 20 June 1991, and issued 20 October 1992. Another US patent (5,358,761), a more complex application also titled "Folded Sheet," was filed in May 1989 and took nearly five and a half years to process.

44. George Wallace McDonald, "Method and Apparatus for Providing Folded Sheets with Stiffeners," US Patent 5,945,195, filed 4 September 1994, and issued 31 August 1999.

45. Albert Soffa and Frederick W. Kulicke, Jr., "Automatic Folding and Binding Machine," US Patent 2,893,297, filed 25 February 1957, and issued 7 July 1959.

46. Sources of information for Walter S. Sachs include the U.S. Public Records Index, an advertisement in a 1949 Haverford College yearbook, various Philadelphia directories, and the Social Security Death Index. He was born in 1930 and died in 1989.

47. Sources of information for Frederick W. Kulicke, Jr., include the 1940 Census, the Social Security Death Index, and Ancestry's Cemetery and Funeral Home Collection. Kulicke died in 2009.

48. Sources of information about Alfred Soffa include the 1940 Census, the Social Security Death Index, and Sally A. Downey, "Albert 'Buddy' Soffa, 84, Technology Pioneer," *Philadelphia Inquirer*, 12 April 2005, B09. Soffa died in 2005.

49. Anders Oswald Palm, "Foldable Sheet," US Patent 2,525,937, filed 29 June 1948, and issued 17 October 1950. Palm, who assigned half his rights to fellow Stockholm residents Nils Gustaf Hoglander and Torsten Frans Moberg, had filed a similar patent in Sweden on 18 March 1947.

50. Patents not previously referenced are Irving Sheroff and Howard W. Berwanger, "Sheet-Folding Machine," US Patent 3,753,558, filed 30 December 1970, and issued 21 August 1973; Stephan R. W. Muth and Guenther Vollath, "Sheet Folding Method and Apparatus," US Patent 4,826,212, filed 10 September 1987, and issued 2 May 1989; Stephan R. W. Muth and Guenther Vollath, "Sheet Folding Method and Apparatus," US Patent 4,917,405, filed 28 April 1989, and issued 17 April 1990; Alfred Vogtländer and Elsbeth Vogtländer, "Apparatus for Folding a Foldable Sheet," US Patent 4,636,192, filed 3 September 1985, and issued 13 January 1987; and Derek Dacey, "Foldable Product with Fold Lines That Are Partly Provided with Creases and Partly by Lines of Perforations," US Patent 8,414,300, filed 14 August 2002, and issued 9 April 2013.

51. PopOut website, accessed 2 August 2015, http://popoutproducts.com/about-us/. The firm is more commonly known in the United States as PopOut Products, perhaps to avoid confusion with an older American firm, Compass Maps of Modesto, California.

52. His "correct" name is Stephan Muth Van Dam. Stephan Van Dam, email correspondence, 31 July 2015.

53. The earlier patent is Stephan R. W. Muth, "Sheet Folding Method and Product," US Patent 4,502,711, filed 6 July 1983, and issued 5 March 1985. Also see Stephan Van Dam, "Mapmaker, Make Me a Map" [interview], in *The Education of a Design Entrepreneur*, ed. Steven Heller, 48–53 (New York: Allworth Press, 2002).

CHAPTER 5

World Views

Because John Snyder had questioned the usefulness of patenting map projections, I approached this chapter gingerly. John developed several of his own projections and was the federal government's map projection guru from the late 1970s through the mid-1990s. He not only advised on geometric distortion in maps and satellite imagery but also authored *Flattening the Earth: Two Thousand Years of Map Projections*, the definitive history of the subject.[1] "Because there are so many freely available projections equal to or better than those patented," he wrote, "the protection sometimes insures the dormancy of the proposal, contrary to the inventor's dreams." Most would-be patent applicants were no less apprehensive: the *Bibliography of Map Projections*, a US Geological Survey (USGS) report that John compiled with geographer Harry Steward, lists 2551 articles and other publications, only 14 of which are patents.[2]

Prudent skepticism drove me to search for other patented projections. My quest might have been easier if the US Patent Classification had a specific category for map projections. Although none of the several thousand class/subclass categories encompasses all 14 patents, three categories collectively cover all but one of them, as either the principal category or a cross-reference.[3] "Printed matter/Maps" (283/34, 7 patents), the marginally more common category, also includes most of the patents discussed earlier in this book, whereas "Education and demonstration/Geography—Map or terrain model" (434/150, 5 patents), the title of which encompasses

© Mark Monmonier 2017
Mark Monmonier, *Patents and Cartographic Inventions*,
Palgrave Studies in the History of Science and Technology,
DOI 10.1007/978-3-319-51040-8_5

most map projections in atlases and geography textbooks, seems moderately ambiguous insofar as all 14 patented projections listed in the USGS bibliography could serve as instructional aids. By contrast, the wordier rubric "Education and demonstration/Geography—Terrestrial globe or accessory therefor, having plural planar or curved surfaces (e.g., flat or frustoconical surfaces, etc.)" (434/135, 5 patents) is a reliably specific description when a whole-world map printed on paper can be cut out and assembled into a three-dimensional object that approximates a sphere. Textbooks have another name, *polyhedron*, for an assemblage of flat facets that can (but need not) mimic a globe.

Because no other patents in my original (283/34–35) database were map projections, I turned to the two promising Education and demonstration subclasses and looked at all patents filed before 1990, a cutoff date chosen to ensure comparability with Snyder's list. I mean *looked* quite literally: because patent drawings capture the essence of an invention, Supreme Court Justice Potter Stewart's oft-quoted standard for obscenity—"I know it when I see it"—is both appropriate and workable as a screening strategy.[4] It's also more efficient than a text search based on *map projection*, which fingered hundreds of mostly post-1990 patents for inventions that merely referred to an existing map projection or cited a book or article with the term in its title. In fact, visual inspection of the 74 pre-1990 patents uncovered by a text search of Google Patents found no plausibly germane patent not already cited by Snyder and Steward.

By contrast, a visual canvass of the two aforementioned Education and demonstration categories identified an additional seven patents, ranging from the irrefutably relevant to the clearly questionable—questionable because the inventor seemed focused more on globes than on flat maps. Although I could have drawn an arguably appropriate line between a globe and a map projection, whatever criteria I used would have purged at least a few patented projections clearly sanctioned by Snyder. Inclusiveness won out.

The oldest patent in my expanded collection of 21 inventions underscores the problem. In 1876 New York City resident J. Marcus Boorman paid $15 to file an application titled "Geometrical Solids for Mapping." His claims referred to solid objects (polyhedra) bounded by 15, 22, 23, 24, or 37 faces, for which he coined names like *quindecahedron* (15 sides), *tricosahedron* (23 sides), and *heptriacontahedron* (37 sides). Each polyhedron consisted of two to five different kinds of polygons—mostly pentagons and hexagons but not always equal-sided. Snyder also mentioned a

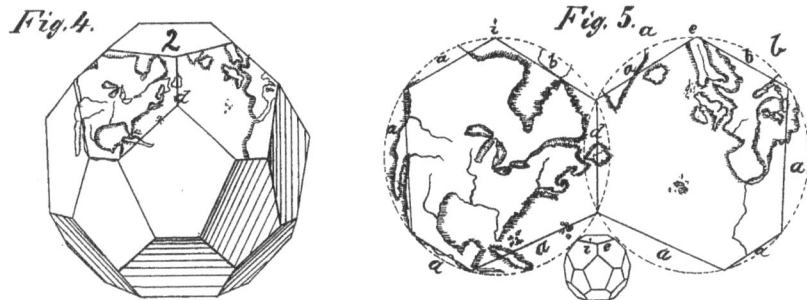

Fig. 5.1 J. Marcus Boorman embellished 2 of his 11 drawings with cartographic features. Only two facets of the chosen solid, a docosahedron (22 sides), were fully mapped. The *left image* shows their positions, and the *right image* provides a more detailed view (US Patent 185,889; 1877)

solid of 32 sides, but neither the patent's wording nor its drawings reflects this configuration.[5] That only two of Boorman's 11 drawings depict a mapping of the globe's coastlines onto one of his polyhedra (Fig. 5.1) highlights his primary intent of "dividing the surface of a sphere into areas for equal maps as little distorted as may be." Arguing that all of his solids were "capable of being inscribed in a sphere," he sidestepped the formal mathematics for projecting a spherical facet onto a plane as well as the strategic anchoring of at least one facet to the equator, one of the poles, or a carefully chosen reference meridian. Neither the conventional grid of meridians and parallels nor the planet's landmasses seemed to matter.

The vagueness of Boorman's procedure undermines both the novelty and the utility of his invention. Snyder noted that polyhedral globes had been pioneered by the distinguished painter and printmaker Albrecht Dürer in 1538 and revived in the early nineteenth century. Although Boorman was apparently the first American to patent geometric solids for mapping, Snyder reported earlier references to polyhedral maps as well as an 1851 British patent that included an icosahedral (20-sided) configuration.[6]

Boorman's claim to having invented "a system of new and useful geometrical solids ... specifically adapted for illustrating the science of solid geometry and mathematics as applied to mapping" met immediate resistance from a skeptical patent examiner who questioned the utility of what he labeled a "curious, probably novel" system. In a one-page rejection that didn't challenge wording or deny specific claims, the examiner contended that "to use a set of irregular figures like these ... to represent the rotundity of the earth for mapping resembles a very forced and unnatural reach after a utility that appears to be inherently wanting."[7] He might have added mind-boggling.

Within a week Boorman's attorney attacked the strict standard of utility of an examiner who, perhaps too generously, had acknowledged the invention's probable novelty. "How useful must an invention be to qualify for a patent?" the attorney asked. Citing a textbook on patenting and a pair of decisions by the Commissioner of Patents—a cogent appeal to higher authority should the case go to court—he argued that an "apparently trifling" degree of utility would not preclude a patent if the invention performed as promised and was not "contrary to sound morals." Dismissing the reason for outright rejection as "mere opinion," he asked the examiner "to confine his action within his legitimate sphere and to pass the case … unless he should be able to show a lack of novelty in said invention." Boorman's attorney, whose letterhead identified him as editor and publisher of "James A. Whitney's Quarterly News-Letter and Patent Law Reporter, A Periodical Devoted to the Interests of Inventors and Patentees," had little tolerance for overreaching rejections.[8]

Intimidated perhaps but hardly humbled, the examiner replied with a more conventionally fussy assessment in which "various amendments [were] necessary to put the application into proper shape." For instance, the physical model Boorman had included did not follow his written and graphic descriptions, and should not, in any event, have been mentioned in the specifications. Particularly problematic were words like *heptriacontahedron*, words not found in dictionaries or mathematics textbooks, words meaningless in a patent claim. Because of Boorman's blatant fondness for coined terms, it is impossible to tell whether the examiner's rejection of "hexagens" and "poly-gens" as unacceptable spellings of *hexagons* and *polygons* might have been a mere (or deliberate) misreading of the cursive *e* in his longhand application—in 1876 most offices lacked typewriters. Obscure terminology was part of the examiner's larger complaint that "many of the sentences … are unduly and objectionably extended and are open to the charge of prolixity." Moreover, all but one of the claims were "too theoretical in subject matter, a mere mathematical figure being regarded as in its nature unpatentable." However novel and potentially useful, Boorman's application could not be approved without "extensive amendment."[9]

Attorney Whitney replied by accepting specific recommendations while defending the application's wordiness. "The exceptional character of this invention should be kept in mind," he argued. "The inventor has taken his own way of describing an invention notable for its originality and which another would hardly describe as well without losing a portion at least of his ideas"—call this the brilliance and novelty defense. Asserting that Boorman had not only invented a geometrical figure but "applied [it] to a practical industrial purpose," Whitney challenged the examiner to propose "other forms of claims covering the same ground"—akin to insisting that the Patent Office rewrite a faulty manuscript.[10] The ploy apparently worked because the case file includes no further protest from the examiner. Boorman paid his additional $20, and in early January 1877, little more than five months after submitting his application, he had his patent.[11]

Boorman's likely motive was achievement, not money—if getting one's name in the *Official Gazette* qualifies as achievement. Psychologist David McClelland, in his 1961 classic *The Achieving Society*, argued that much of human behavior can be attributed to three basic needs: achievement, affiliation, and power.[12] McClelland was fascinated with the need for achievement as reflected in the focused pursuit of prizes, successful business ventures, or the recognition (real or imagined) presumed to follow publication of a book, a poem, a scientific article, or a clever patent. Money is not the only symbol of achievement, he observed; diverse accolades can confer a sense of accomplishment. And if an inventor's goal is clearly not economic, it's probably the need for recognition shared with amateur athletes, deer hunters, summer stock actors, people who enter puzzle contests, and most university professors—a basic human agenda that conflates being noticed and being loved.

Boorman's obscure sense of his invention's usefulness suggests his goal included naming rights for a new set of geometric solids, an obvious achievement for a recreational mathematician. Though city directories list him as a clerk and later as a lawyer—it is not clear where or whether he obtained a law degree[13]—Boorman's interest in mathematics is apparent in the several articles he published in *The Mathematical Magazine: A Journal of Elementary Mathematics*, which described him in its list of contributors as a "Consultative Mechanician, and Attorney and Counselor at Law."[14] In an era when professional training and credentials were often informal, a *consultative mechanician* was probably a free-lance machinist or mechanical engineer. This interpretation is consistent with Boorman's receiving US patents for a hay raker in 1868, a rotary steam engine in 1869, and a hay loader in 1870—successes that probably encouraged him to patent his geometric solids a half decade later.[15] Although it is not clear that he profited significantly from any of these mechanical inventions, he filed patent applications in England, where the London weekly *Engineering* noted his steam engine's "fearful and wonderful appearance."[16] Boorman received one more US patent, in 1894, for a small sailing vessel with a hollow, water-tight outrigger.[17] As before, his attorney was James A. Whitney, whose office, a few doors away on Broadway, was conveniently close for a knowledgeable enabler.[18]

Although Boorman's map projection patent most likely reflects a bright, legally adept mathematical hobbyist eager for recognition, fame is less apparent an explanation for the patent granted to William Wilson, of Edinburgh, Scotland, for a "Geographical Map, Globe, and Other Geographical Appliance."[19] His drawings (Fig. 5.2) describe a set of globe gores, 45° of longitude wide, which can be assembled into a polyhedral globe with six belts, each covering 30° of latitude. The uppermost and lower-left drawings show how the eight gores, which are connected near the equator, can be formed into a cylinder by joining flaps (labeled *e*) on the left and right, and the lower-right drawing describes a spindle designed to secure tabs (labeled *b*) at the top and bottom when the gores are pulled together at the poles. "The map can thus be made up as a globe [as shown in his] Fig. 2 or opened out and used as a flat map at the convenience of the pupil—forming a perfect mental bridge between the usual round globe and the flat map." Although Wilson's intent was clearly pedagogic, I have yet to confirm that he actively marketed the invention as a teaching aid. Even so, his belief in its usefulness is apparent in patents secured in Britain and the United States as well as his 1939 donation of "geographical models and maps specially designed for educational purposes" to the library of the Royal Scottish Geographical Society.[20]

If Snyder knew of Wilson's invention, he probably ignored it because the patent fails to anchor the 48-sided polyhedron to a conventional geographic framework and says nothing about the projection's coordinate geometry. Would Snyder have mentioned the invention in *Flattening the Earth* had Wilson sketched in a continental coastline or two, as Boorman had done? Perhaps. Although the Scottish pedagogue bypassed the formal mathematics, his verbal narrative offered a clear description of facets with straight-line parallels perpendicular to (and evenly spaced along) their central meridian, and straight-line meridians evenly spaced along the parallels. Adding simple cartographic features to one of two of the facets could have made the apparatus look more like a flat map and less like a collapsible globe.

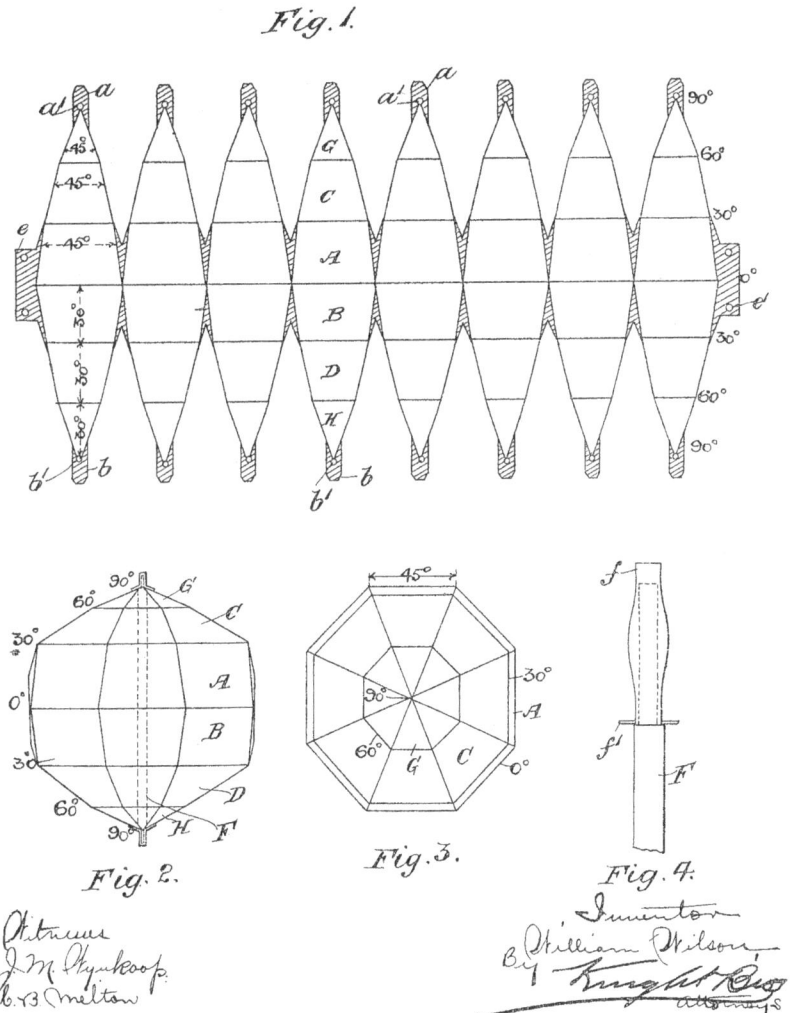

W. WILSON.

GEOGRAPHICAL MAP, GLOBE, AND OTHER GEOGRAPHICAL APPLIANCE.

APPLICATION FILED JULY 28, 1909.

944,248.

Patented Dec. 21, 1909.

Fig. 1.

Fig. 2.

Fig. 3.

Fig. 4.

Fig. 5.2 William Wilson's drawing sheet describes a flat map that converts readily into a polyhedral globe (US Patent 944,248; 1909)

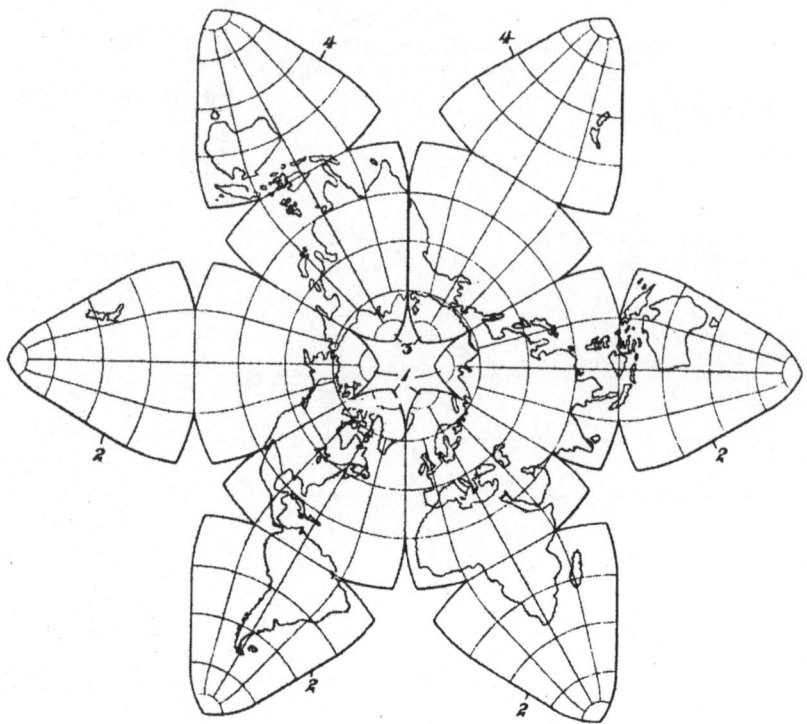

Fig. 5.3   The single drawing included with Bernard Cahill's first patent. The two uppermost lobes, which are redundant, touch along a darker section boundary, across which the western shoreline of North America appears to continue as the shoreline of South Asia (US Patent 1,054,276; 1913)

By contrast, the distinction between a globe and a map projection is more readily apparent in the two patents filed in 1912 and 1913 by San Francisco architect Bernard J. S. Cahill. His earlier patent, titled "Map of the World," suggests a globe divided like an unpeeled orange into six lobes, identical in size and shape (Fig. 5.3).[21] Each lobe is divided further into geometrically identical northern and southern sections, partly joined along a straight line at the equator. Although the similarity to orange peels might be striking at first glance, close inspection reveals that the 12 sections could not be pressed downward neatly onto a globe without stretching, tearing, and awkward overlap. Notice the two Australias, each on a different lobe; the two New Zealands, each dominating the southern portion of two other lobes; and a perplexing duplication in the northern hemisphere, where the west coast of North America meets northeast Asia twice, across two Bering Straits. There's a point to this duplication, which not only provides at least one fully intact, uninterrupted rendering of each continental landmass—except for Antarctica, conveniently absent from the patent drawing—but also affords minimally uninterrupted versions of the North Atlantic and the North Pacific Oceans.[22]

Cahill achieved this cartographic sleight of hand by adopting a graticule (network of meridians and parallels) with grid lines 22.5° apart, laying out lobes that cover 90° of longitude, and interrupting the lobes and sections with short 22.5° cuts along bounding meridians and the equator. The illusion of a seamless map is betrayed by the slightly thicker and noticeably darker meridional boundary running directly upward from the map's center. To its right—if you resist the temptation to rotate the page—is a slice of western North America bounded on the east by a meridian (112.5°W) extending poleward from Baja California. And to its left is a major portion of Asia bounded on west by a meridian (67.5°E) running northward from the coast of Pakistan near its border with India.

Cahill's uninterrupted continents and northern oceans could have been inspired by the longitudinally extended Mercator grids used in the late nineteenth century to describe shipping routes reaching eastward and westward toward dual Australias near the left and the right edges of the map.[23] In much the same way that a cylindrical projection like the Mercator has room for repeat coverage on its left or right sides, a conic projection covering the whole world has space for duplicative content because its projected parallels are not full circles. But unlike a cylindrical projection, a conic projection cannot be extended indefinitely. Cahill's patent describes a conic projection designed to cover the entire world with four lobes requiring only 240°, thereby leaving 120° for the two repeated lobes.[24]

In an article published in the *Scottish Geographical Magazine* in 1909, three years before he filed his patent application, Cahill described a protracted trial-and-error process begun well before the April 1906 San Francisco earthquake and fire destroyed his home, belongings, and experimental drawings.[25] An early goal was a scientifically valid alternative to the Mercator projection, notorious for exaggerating areas in upper latitudes and used widely—and inappropriately—for situations unrelated to navigation, its intended purpose. For most purposes the "demands" of land were more important than those of water, and "looking down at the Pole" would serve these demands better than looking "sideways at the Equator." In making experimental maps, he sought an "axial meridian" for each continent. He regretted having to separate Australia and New Zealand, "but as the New Zealanders themselves much prefer it this way, this feature of the map need hardly call for our regret."

Leery of a Patent Office rule that precluded patenting an invention described in print more than two years prior to filing, Cahill disclosed his 1909 article, which "does not contain a description of the invention as finally completed"—true but mildly contradictory insofar as the last of the article's 14 illustrations explaining the evolution of his thinking is captioned "The Completed Map."[26] Like the subsequent patent's single map, its 1909 predecessor shows four regular and two repeating lobes, but one of the extra cycles repeats the Americas, not Australia, and the other oddly suppresses an encore of eastern Asia and New Zealand. Cahill continued to tinker with his invention over the next two decades, and ultimately produced versions designed to preserve angles, relative area, or great circle routes.[27]

Cahill's patent application was brief and heavy on hype, proclaiming the invention's superiority to the Mercator projection and touting the "great advantages" of a map with multiple lobes and sections that "can be folded upon one another in a very compact form, one-twelfth the size of the original map"—a weak justification in the eyes of a patent examiner who saw no advantage "unless it is the ability to fold" and rejected the application for sacrificing novelty for usefulness.[28] In reply, Cahill's attorney dismissed folding as "only an incidental advantage" and emphasized the more accurate treatment of landmasses.[29] He concluded by referring to the peeling of an orange, which can be pressed flat with less distortion if first cut into eight sections—a convincing rebuttal rewarded seven weeks later, when his application was officially "allowed."[30]

Cahill's second patent, filed less than seven months later, emphasized the eight-section advantage. Although a tiny globe at the top of the single drawing sheet justified the title "Geographical Globe," the principal image was eight equilateral spherical triangles flattened into an assemblage later dubbed the "butterfly" projection (Fig. 5.4).[31] John Snyder identified this configuration as an octant map, a format pioneered in the sixteenth century by Leonardo da Vinci.[32] In later versions (not patented), equilateral triangles with straight sides that fit together neatly replaced the curved octants.[33] Snyder acknowledged Cahill's three-decade campaign to pro-

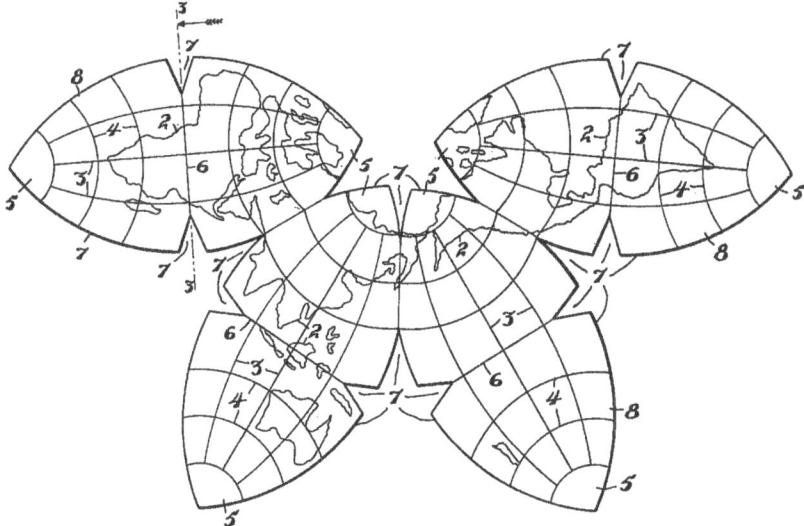

Fig. 5.4 Bernard Cahill's second patent featured a map with eight sections spread out like a butterfly pinned to a board (US Patent 1,081,207; 1913)

mote his projections but was unimpressed with the results, calling the projection "almost unused."

Cahill's rediscovery of the octant map reflects significant earlier experience in the applied visual arts. Born in England in 1867, he studied at the University of London and the South Kensington School of Art, moved to California in 1888, and worked as a draftsman for three years before becoming a practicing architect. His entry in *Who's Who on the Pacific Coast* for 1913 lists numerous accomplishments as an architect, city planner, and contributor to professional journals, in addition to "inventor of a new projection for land maps of the world without exaggeration or distortion to supersede Mercator's" and "Pres. And Mgr., 'Cahill's Land Map of the World' (inc.)"—which might account for his seeking a patent. Incorporating the business, a less complicated process, is further evidence of his deep but not widely shared belief in the octant worldview.[34]

Though the business was never more than a sideline, Cahill promoted his butterfly projection energetically but intermittently over the next three decades. Articles in the *Monthly Weather Review* in 1929 and 1940 and in *Architect and Engineer* in 1939 touted the benefits of partitioning the globe into eight sections, for which regionally specific projections could minimize distortion of—take your choice—area, angles, or great-circle direction.[35] Cahill died in 1944, but his name and octant concept survive in the Cahill-Keyes world map, a tweaked version presented in 1975 by political scientist Gene Keyes, who maintains an Internet shrine of sort to Cahill.[36] In addition to a biography and bibliography of the octant apostle, the B. J. S. Cahill Online Resource describes architect R. Buckminster Fuller's Dymaxion projection as not only awkward but also inferior to Cahill's butterfly.[37]

While Keyes stopped short of denouncing Fuller as a copycat, Cahill was a likely inspiration for several inventors who sought recognition at the Patent Office for mapping the globe onto a polyhedron. Figure 5.5 compares graphic excerpts from four patents filed between 1937 and 1944 for polyhedral map projections that could cover a solid with triangular, square, or pentagonal facets. With more than eight facets, all could control geometric distortion better than Cahill's butterfly map if users willingly overlooked disconnected continents and severed seas.

For example, James Addison Smith, an Internal Revenue Service agent in Seattle, mapped the planet onto a dodecahedron (12 sides, all pentagons) with a facet centered on each pole and a single pentagon encompassing most of North and Central America (Fig. 5.5, upper left). His patent, filed in 1937 and granted a year and a half later, is titled simply "Globe."[38] All three of its claims begin with "An approximate globe comprising a collapsible frame," which suggests his goal was not two-dimensional. Perhaps because one of his patent's four figures depicted three pentagons connected together like a flat map, Snyder mentioned it along with several other polyhedral projections, none of them patented but all possibly inspired by Cahill.[39] Forty-nine years old when he filed his application, Smith had four years of college, but apparently no prior experience with drafting, mapping, or the Patent Office. Unlike Cahill, whatever effort he made to promote his invention left no trace.

By contrast, Joel Crouch, who filed a patent application seven years after Smith for an "Icosahedral Map," included a fully opened-out flat map of the world (Fig. 5.6, above) as well as a pictorial view of its associated globe (Fig. 5.5, upper right), an assemblage of 20 equilateral triangles known as an icosahedron.[40] Although Crouch, like Smith, never promoted his invention, he had appropriate credentials. On a July 1947 questionnaire for faculty members at the Pennsylvania State College, the 47-year-old associate professor of industrial engineering reported bachelor's degrees

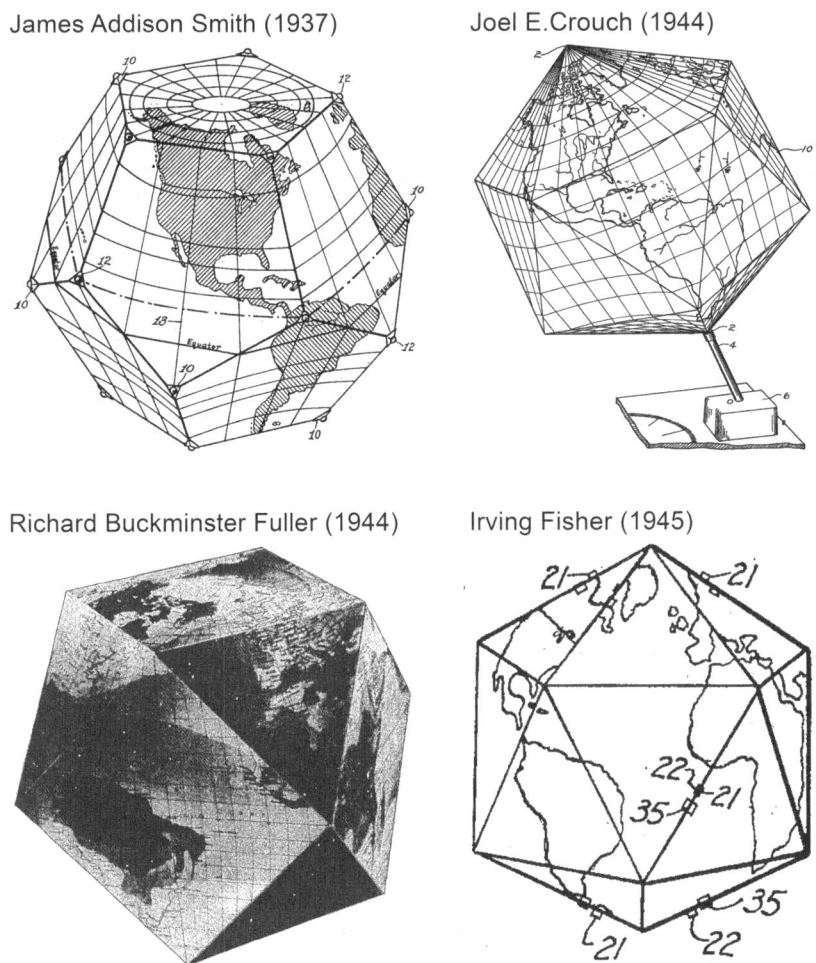

James Addison Smith (1937)

Joel E.Crouch (1944)

Richard Buckminster Fuller (1944)

Irving Fisher (1945)

Fig. 5.5 Excerpts from patents for four polyhedral projections identified by inventor and year of filing. R. Buckminster Fuller's polyhedron was apparently reproduced in his patent from a photograph, not a line drawing (*Upper left*: US Patent 2,153,053; 1939. *Upper right*: US Patent 2,424,601; 1947. *Lower left*: US Patent 2,393,676; 1946. *Lower right*: US Patent 2,436,860; 1948)

Joel E. Crouch (1944)

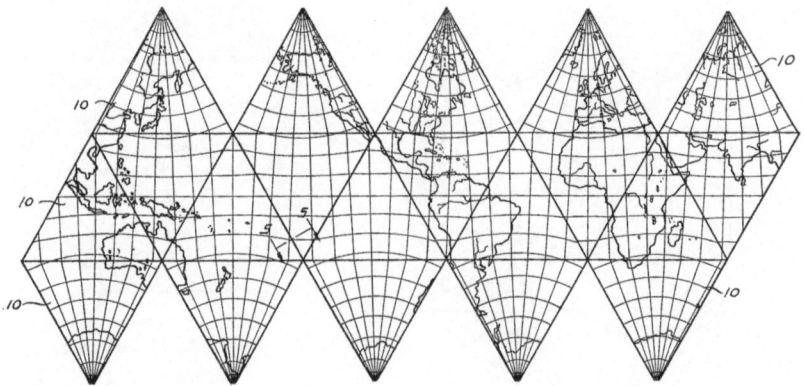

Irving Fisher (1945)

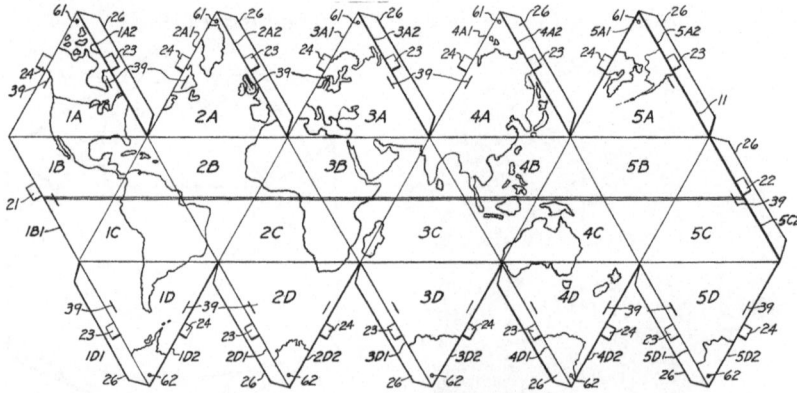

Fig. 5.6 Flattened views of the whole-world maps devised by Joel Crouch and Irving Fisher, who used assemblages of 20 equilateral triangles. Year of filing is shown in parentheses after the inventor's name (*Above*: US Patent 2,424,601; 1947. *Below*: US Patent 2,436,860; 1948)

in mathematics (1920) and civil engineering (1923) as well as a master's in industrial engineering (1941).[41] He also listed a single hobby, celestial navigation, which explains his only other patent, titled "Astronomical Device" and filed the same day, 25 January 1944, as his application for the icosahedral map shown here.[42] The link between the two inventions is functional as well as temporal insofar as the Astronomical Device consists of a 20-sided celestial globe that surrounds a much smaller 20-sided terrestrial globe. The two globes share the same axis, but the celestial globe is attached to a sleeve that allows it to rotate independently. Points on the outer icosahedron represent constellations, some labeled, and lines on the inner icosahedron represent coastlines—a clever idea well hidden in the *Official Gazette* summary.

Crouch's failure to promote his invention with even a short note in a scientific journal probably accounts for its absence from John Snyder's impressively thorough history and bibliography, compiled more than two decades before Google Patents—which begs the question: How did Snyder learn of Smith's invention? I doubt that Snyder would have dismissed Crouch's map as only a geometrical solid with a cartographic skin. Indeed, the Icosahedral Map's dual role as a polyhedral globe is not materially different from the similar duality of Irving Fisher's "Global Map" (Fig. 5.5, lower right; Fig. 5.6, below), also based on 20 equilateral triangles.[43] Indeed, the drawing for Fisher's flat map includes reference numbers for tabs (23 and 24) and slots (39) to be used for fasteners when the printed sheet is cut, folded, and assembled into a three-dimensional model of the world. A key difference is publicity: unlike Crouch, Fisher presented his projection to an audience of geographers and mapmakers in a short article published in the *Geographical Review* a year and a half before he filed with the Patent Office.[44] In addition, he not only arranged for the sale of a printed map that could be folded into a globe but also convinced a children's magazine to run a four-page article with a cut-out icosahedral globe for readers to assemble.[45]

Fisher's emergence as a cartographic inventor is more readily explained: a prominent academic economist, he was skilled in mathematics and had a long history of devising mechanical gadgets, some patented and one highly profitable. A card index patented in 1912, when he was 45 years old, became the foundation for Index Visible, Inc., a record system business that earned him several million dollars, most of which he lost after the stock market crashed in 1929.[46] Less lucrative inventions include a bed for tuberculosis patients, a three-legged folding seat, and a mechanical device for calculating a meal's nutritional balance. A professor of economics at Yale, Fisher was elected president of the American Economics Association in 1918. He was also a prolific author, whose oeuvre includes 28 books, beginning with *Elements of Geometry* (1896) and ending with *World Maps and Globes* (1944), co-authored with Osborn Maitland Miller, a projection expert at the American Geographical Society, in New York.[47]

His biographer Robert Loring Allen, in a final chapter titled "Moving into the Shadows," described how Fisher connected with cartography in his late 70s, when World War II had heightened public interest in world maps.[48] Fisher probably met Miller during one of his frequent trips to New York, to participate in corporate board meetings, attend plays, and consult with a physician who specialized in dietary therapy. Miller had recently devised a cylindrical world map that mollified the Mercator projection's extreme distortion of area in higher latitudes.[49] Miller's endeavor, encouraged by the lead geographer at the State Department, might have inspired Fisher to devise a different, no less worthy solution. Unlike Miller, who was content to describe his invention in the academic press, Fisher submitted his "Global Map" to the Patent Office, which issued the patent nearly ten months after his death, at age 80.

Like most attempts to patent a map projection, Fisher's application was approved only after the inventor and his attorney agreed to drop or consolidate claims, reduced in this case from 12 to 7. Oddly, the examiner's single letter of rejection focused not on the map's geometry but on the means for joining together the triangular facets, which seemed insufficiently novel.[50] Perhaps Fisher had too enthusiastically credited Albrecht Dürer with the underlying principle behind foldable icosahedron maps, which (the application noted) "did not come into practical use because of the difficulties in converting them from their unfolded form into their folded spherical shape." Fisher's attorney, in asserting his client's inventiveness, emphasized others' failed attempts to implement Dürer's idea in a tortuous sentence only a patenting professional can appreciate: "Anything which was not obvious to skilled scientists who tackled this problem in the past," he argued, "must therefore be considered to be unobvious to, and as not within the ordinary expectable skill of the ordinary skilled workers in the art."[51] The response to the examiner's concerns also included numerous changes in wording and spelling, such as adding the umlaut to *Dürer*, replacing "globe" with "globe-like form," and changing the second *o* in *icosahedron* to an *a* and then changing it back. A dance of sort but necessary to win approval.

No less odd is the failure of the examiner who vetted Fisher's invention to notice similarities with Crouch's patent, distinguished largely by a different method of assembling the facets (Fig. 5.6). Both Crouch and Fisher used great circles to partition the globe into 20 equilateral spherical triangles, which they then projected into a network of planar triangles by gnomonic projection, which flattens all great circles into straight lines.

This similarity notwithstanding, Crouch's patent, filed just 13 months before Fisher's, had not been granted when Fisher filed. Had someone at the Patent Office noticed the resemblance, the Interference Division would probably have intervened.

John Snyder had little to say about Fisher's projection, treated in a single paragraph under the subheading "More Polyhedral Globes" along with R. Buckminster Fuller's Dymaxion map.[52] Fuller had used *dymaxion*, coined from *dy*namic, *max*imum, and tens*ion*, for several earlier inventions, including structures and an automobile. For the map he applied the term to a polyhedron more generally called a cuboctahedron because it combines all the elements of a cube and an octahedron: six squares and eight equilateral triangles. Careful examination of the perspective image from his patent (Fig. 5.5, lower left) reveals a solid bounded on the top and bottom by squares connected at their corners to four additional squares, each rotated 45°. Eight equilateral triangles that share sides with these six squares complete the solid. All perimeters of squares and triangles represent portions of great circles and share the same scale. Unlike Fisher's map, on which gnomonic projection increased distortion outward from the center of each triangle, Fuller's map increased distortion inward from the facet's periphery. Even so, the overall distortion of landmasses was at least marginally lower.

Fuller had been intrigued by the idea of "one-world island in one-world ocean" since the 1930s, and included a sketch map similar to artwork in his patent application on the end papers of his 1936 book *Nine Chains to the Moon*.[53] He no doubt sought patent protection because he had seven patents to his name and clearly understood the patent system when he filed for an invention enigmatically titled "Cartography" in February 1944.[54] He was an accomplished inventor, and filing patents is what he did. Fuller was also an accomplished self-promoter, who convinced *Life* magazine to run a 15-page photographic essay titled "R. Buckminster Fuller's Dymaxion World" in its 1 March 1943 issue. Included were a cut-out world map and instructions for its assembly.[55] In 1940, when he was a science and technology consultant for *Fortune*, the magazine published "World Energy: A Map by R. Buckminster Fuller," framed of course on the Dymaxion projection.[56]

Why the Patent Office accepted a photograph of a three-dimensional model (Fig. 5.5, lower left) instead of a conventional patent drawing, with crisp lines drafted in India ink, is a mystery.[57] The image reproduces poorly, here as well as in the published patent, as does the photographic copy of an unfolded flat map (Fig. 5.7, above), supplemented by a tradi-

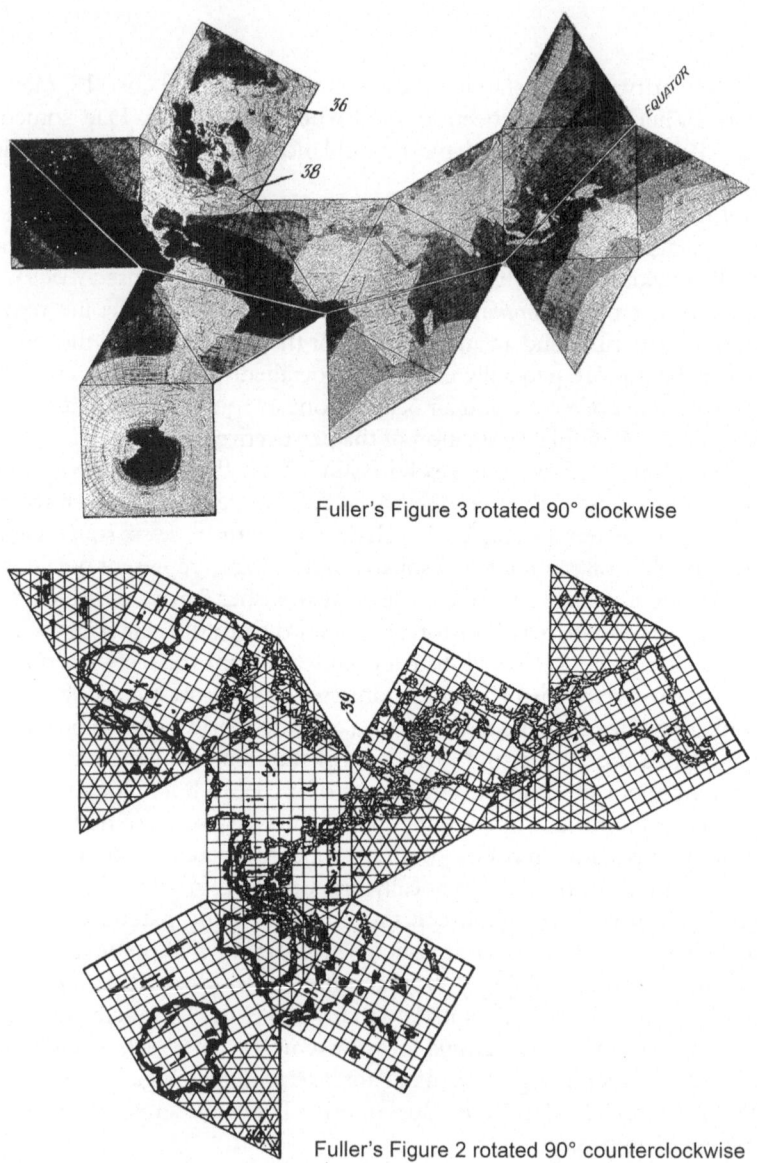

Fuller's Figure 3 rotated 90° clockwise

Fuller's Figure 2 rotated 90° counterclockwise

Fig. 5.7 Fuller's patent shows different ways of fitting a cuboctohedron to Earth's landmasses. Both images were rotated 90° to fit them into the same illustration, for ready comparison. The upper map anchors the polyhedron to the poles, whereas the lower map offsets the North Pole (numbered 39) to further minimize interruption of landmasses. The numbers 36 and 38 on the upper image merely pointed out the grid of meridians and parallels on square (36) and triangular (38) facets (US Patent 2,393,676; 1946)

tional black-and-white drawing with coastlines rendered in a wide zig-zag symbol—coastlines largely illegible until I used Photoshop to thicken all lines (Fig. 5.7, below). In the case file at the National Archives, a patent examiner complained about a blurry version of this latter map, which was redrawn after the related claims were approved.[58] The examiner also complained that Fuller used *dymaxion*, not found in any dictionary, as if the term were common knowledge.[59]

Why two separate unfolded worldviews? Close inspection of Fig. 5.7 reveals two different embodiments of Fuller's invention. For the upper image he placed the poles at the centers of opposite squares, for an uninterrupted view of Antarctica and an equator represented by a chain of straight-line segments connecting opposite corners of the other four squares. By contrast, on the lower map "the north pole is located in an arbitrary position … selected with [a] view to having the land areas so placed as to eliminate land sinuses"—unfolding the cuboctahedron could introduce unwelcome gaps (sinuses) when adjoining triangles were separated. Fuller experimented with other arrangements and orientations of the polyhedron's facets to accommodate labels like "East by Steam to the Orient via Suez" and "Stratosphere Strategic," with Europe at the center, surrounded by Africa and Asia.[60]

Fuller continued to tweak his invention, most noticeably by abandoning the 14-sided cuboctahedron for a 20-sided regular icosahedron. Unlike Fisher's gnomonic icosahedral map, anchored at the poles, Fuller kept landmasses intact by avoiding the poles and dividing two of the triangles into three non-equilateral triangles and one quadrilateral.[61] Instead of patenting the new map, he chose instead to copyright a printed version and trademark the name Dymaxion.

Integrity of the world's land masses became the dominant focus. Mapmakers who licensed the concept often suppressed the network of triangular edges and omitted the oceans for a clean view of the land. Paul Knox and Sallie Marston, who adopted the Dymaxion projection for thematic maps in their popular introductory world geography textbook, even omitted Antarctica, inconvenient and often irrelevant.[62]

Between 1974 and 1987, three foreign inventors filed patent applications for polyhedral map projections, which were approved by the Patent Office but overlooked by Snyder. Jean Thorel and Henry Dufour were Frenchmen, and Su Hi Wang was a resident of Taiwan. All three projections (Fig. 5.8) are sufficiently distinctive to merit patent protection, but I found no record of their commercial development. Thorel's

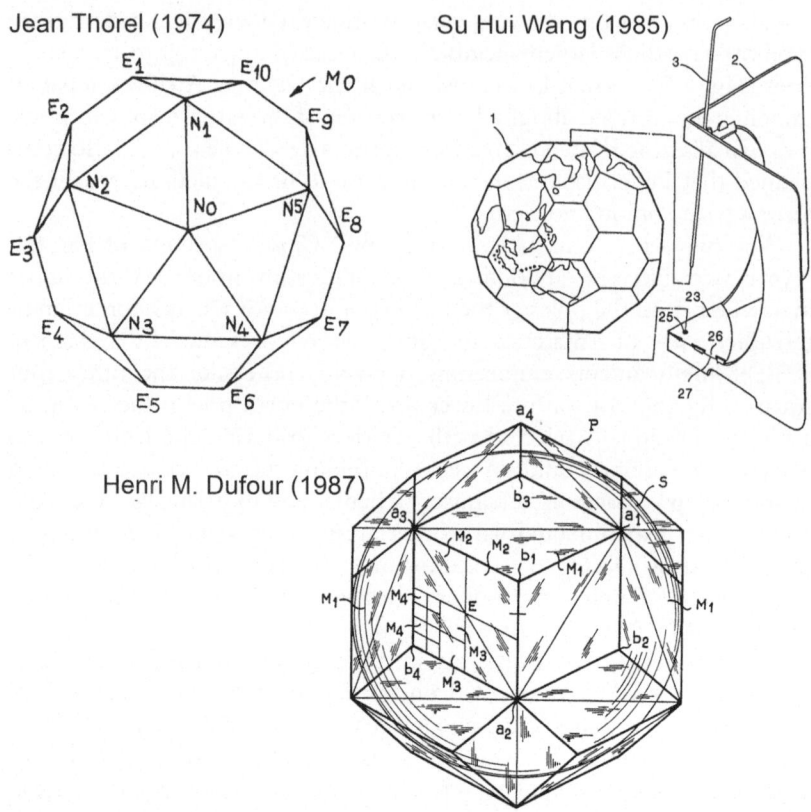

Jean Thorel (1974)   Su Hui Wang (1985)

Henri M. Dufour (1987)

Fig. 5.8 Excerpts from patents for three polyhedral projections identified by inventor and year of filing. None of the three inventors were US residents (*Upper left*: US Patent 3,868,781; 1975. *Upper right*: US Patent 4,620,842; 1986. *Bottom*: US Patent 4,773,861; 1988)

images show a polyhedron consisting of triangles and trapezoids, none embellished with coastlines or a graticule.[63] Although his patent is titled "Polygnomonic Map of the World Comprising Two Hemispheres," its claims focus solely on the geometric framework. Wang's images, one of which includes generalized coastlines, describe a polyhedron of 32 pentagons and hexagons with tabs and slots for assembly into a globe that can be mounted on a pedestal, as shown.[64] The goal is largely pedagogic: a map of the world assembled by hand, piece by piece, to engage and "better enlighten the pupils of global geography." By contrast, Dufour's patent describes "a semi-regular polyhedron having thirty equal faces of lozenge shape," which can be subdivided into successive steps, leading ultimately to a spherical grid of millions of small elements for recording elevation or land cover in a geographic database—a spherical version of the hierarchical quadtree and nine-tree data structures (Figs. 2.3 and 2.4) discussed in Chap. 2.[65] Focused on a grid system rather than a world map or a globe, it was assigned to France's national mapping agency, the Institut Geographique National, at which Dufour, apparently a geodesist, was probably an employee or contractor.

In 1977 F. Webster McBryde filed a patent application for a map projection with multiple embodiments, one similar to projections cast on a polyhedron.[66] Not a polyhedral projection per se, his map is configured differently for the northern and southern hemispheres, divided among three and four lobes, respectively, to keep landmasses intact (Fig. 5.9). Each lobe is anchored by a central meridian, plotted as a straight line perpendicular to the equator. Because the central meridians are axes of low distortion, the projection is well suited for thematic maps of population, agriculture, and other land-specific phenomena. Because the parallels are mutually parallel, as on a cylindrical projection, this type of framework is called an interrupted pseudocylindrical projection. Because McBryde's poles are lines, not points, east–west stretching still increases markedly with latitude but angular distortion is not as great as if the poles were points.

Others had devised interrupted cylindrical projections, most noticeably John Paul Goode, who used his homolosine equal-area map for the Rand McNally school atlas bearing his name, but only McBryde obtained

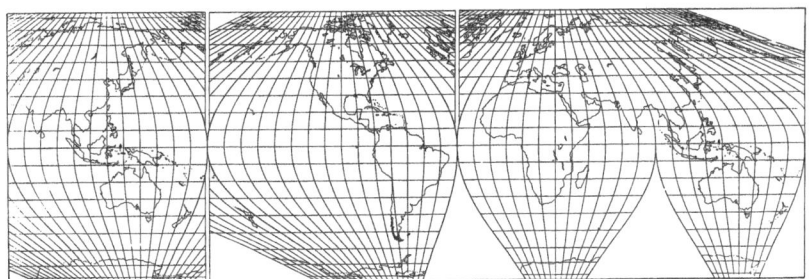

Fig. 5.9 F. Webster McBryde's patented map projection included this interrupted version (US Patent 4,315,747; 1983)

a patent, probably because he thought he could license his invention to map publishers. Although he held a PhD in geography from Berkeley and taught geography at Ohio State between 1937 and 1942, McBryde worked mostly as a consultant to federal agencies and nonprofit organizations. In the late 1930s, he drafted and published an interrupted equal-area world map on a projection devised by the renowned German cartographer Max Eckert, and in the late 1940s he collaborated with mathematician Paul Thomas on several flat-polar map projections, developed by fusing existing projections for a more realistic treatment of polar areas.[67] In the late 1970s, he devised new solutions, described graphically and mathematically in his first and only patent, apparently part of a larger effort to publicize (if not monetize) the invention. Particularly enduring is the McBryde S3, a variant interrupted over land to frame a map of Exclusive Economic Zones extending 200 nautical miles outward from the world's coastlines. Snyder referenced several conference papers by McBryde and included an image of the S3 in *Flattening the Earth*.[68] Software applications and handbooks on map projection typically include the McBryde S3.[69]

Oceans and continents provided a focus for another series of composite map projections, presented between 1942 and 1989 by Athelstan Spilhaus, a South Africa-born scientist and inventor respected for contributions to meteorology, oceanography, and anti-submarine warfare.[70] Spilhaus understood the usefulness of world maps constructed as assemblages of the facets of a polyhedron or the lobes of an interrupted pseudocylindrical projection, but he also recognized the shortcomings of great-circle boundaries between these low-distortion components, typically framed by meridians and the equator. Although rotating a polyhedral framework away from the poles, as Fuller had done, could minimize cuts across landmasses, the boundaries between sections did not recognize natural structures like tectonic plates. In a 1983 article titled "World Ocean Maps: The Proper Places to Interrupt," Spilhaus argued for recognizing cartographically that "the ocean has three lobes when spread out"—the Pacific, the Atlantic, and the Indian.[71] Eight years later, in an article co-authored with John Snyder and titled "World Maps with Natural Boundaries," Spilhaus showed how maps interrupted along shorelines could provide seamless representations of both the oceans and the continents.[72]

An appendix to the article describes the evolution of Spilhaus's thinking, starting with a 1942 *Geographical Review* article titled "Maps of the Whole World Ocean."[73] He returned to the problem three decades later, when he sought to incorporate plate boundaries into cartographic experiments, which reminded him of plate tectonics pioneer Alfred Wegener's discovery that the Atlantic coasts of Africa and South America fit together like pieces of a jigsaw puzzle. Realizing that a cartographic jigsaw puzzle could be a marketable teaching aid, Spilhaus filed a patent application in March 1985 for an invention titled "Map Puzzle Having Periodic Tesselated Structure."[74] He received his patent less than two years later, around the time GeoLearning Corporation, in Sheridan, Wyoming, began marketing "Geodyssey, the World Game," the "Spilhaus Geoglyph: a Map of the Oceans and Continents of the World," and similar products.[75] That Spilhaus patented and marketed his map puzzle is not surprising: since 1934 he had patented 12 earlier inventions as diverse as an astronomical clock and ice skates.[76]

Drawings accompanying the patent refer to both a map projection and a series of puzzles created using different configurations of the projection, based on an equilateral tetrahedron (4 sides). Spilhaus was careful to define *tessellated* in the patent's title by noting "the puzzle pieces fill the spaces within the boundary of the particular configuration into which the pieces are arranged without gaps within the interior of the boundaries"— verbose patentese explained graphically by a pair of maps (Fig. 5.10).[77] Each map includes seven puzzle pieces "of arbitrary shape," which can be subdivided further to provide a desired degree of complexity. The map would also have a rectangular frame (not shown in Fig. 5.10), to provide added clues for the novice player. A sentence within each frame indicates a

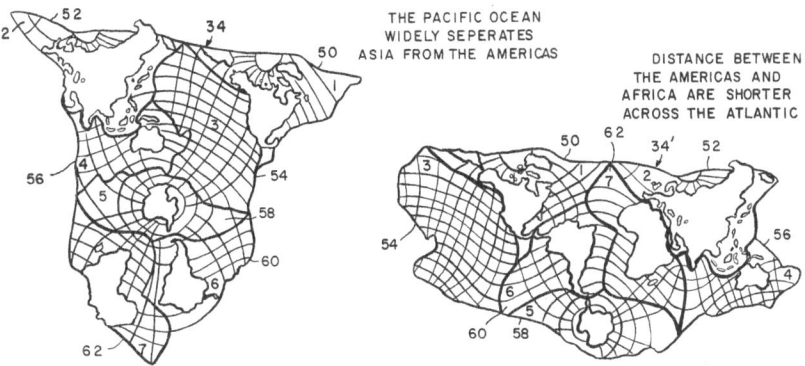

Fig. 5.10 Athelstan Spilhaus's patent for map puzzles included these two configurations of the related map projection (US Patent 4,627,622; 1986)

particular learning point, such as the extreme breadth of the Pacific Ocean (Fig. 5.10, left). That each of the 17 claims begins "A map puzzle ..." makes clear that the map projection, described elsewhere in the scientific literature, is secondary to a game focused on learning about the continents and oceans.

Most whole-world map projections are comparatively straightforward worldviews, uninterrupted by gaps between puzzle pieces, lobes, or poly-hedral facets. This simplicity as well as roots reaching back two millennia explains their relative scarcity as patented inventions—a new uninterrupted projection would not easily meet the standards of novelty and non-obviousness required for a patent much less offer a sufficiently unique and useful geographic framework that a map publisher would pay to license—as John Snyder observed, suitable alternatives were readily available in the public domain. It is thus understandable that my dataset of 21 patented map projections contains only four uninterrupted whole-world map projections, only one of which merited discussion in Snyder's *Flattening the Earth*.

That the other three, represented by excerpts in Fig. 5.11, were listed in the USGS bibliography probably reflects the prominence of the inventor or his employer.[78] Henry de Beaumont was a leading Swiss geographer, who was 69 years old in 1888 when he filed a US patent application for a map with parallels and meridians constructed as circular arcs (Fig. 5.11, top).[79] The projection accords with his opposition to universal time based on the Greenwich meridian: note the straight-line meridian approximately 10° east of Greenwich on his small, highly generalized drawing. I have yet to find a published map cast on this projection.

Chicago draftsman Jules Colas was 51 years old when he filed to patent "a map of the earth having the outline of an oval" and "designed to show the geographic features of the entire earth with approximate correctness" (Fig. 5.11, middle).[80] An amalgamation of four smoothly blended circular arcs formed a bounding oval within which meridians and parallels were laid off evenly. Although the projection preserves neither angles nor area, distortion is lowest along the equator as well as along the central meridian at 90°W, near Chicago. Colas assigned the patent to his employer, Poole Brothers, a prominent cartographic printing firm that rarely, if ever, pro-duced world maps.[81] Because the firm mostly published sectional and local maps, not world atlases, I doubt that Colas's projection was used exten-sively, if at all. This was his only patent.

George Washington Bacon, who created the third patented projection in Fig. 5.11, was a prolific British author, publisher, and mapmaker as

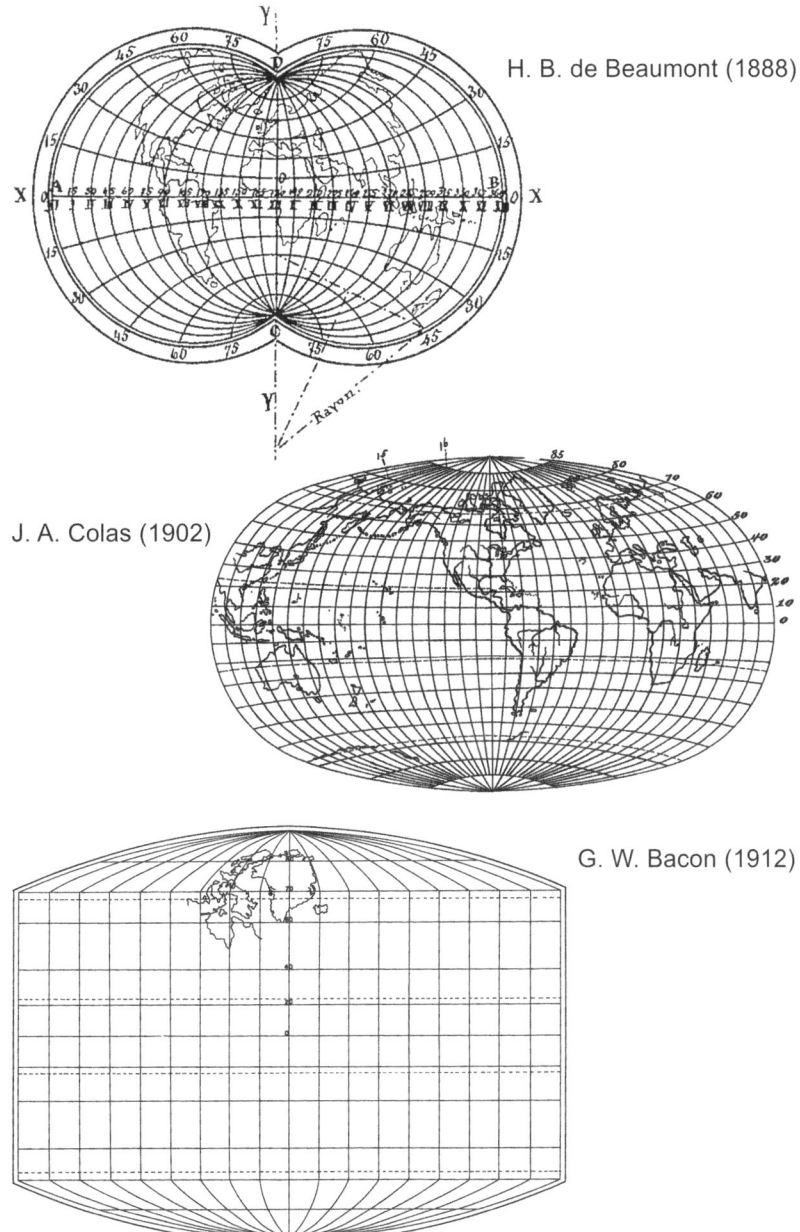

H. B. de Beaumont (1888)

J. A. Colas (1902)

G. W. Bacon (1912)

Fig. 5.11 Excerpts from patents awarded to Henry Bouthillier de Beaumont, Jules A. Colas, and George Washington Bacon. Year of filing is shown in parentheses after the inventor's name (*Top*: US Patent 400,642; 1889. *Middle*: US Patent 752,957, 1904. *Bottom*: US Patent 1,050,596; 1913)

well as a map seller with a shop in central London.[82] Born in Lockport, New York, in 1830, he emerged in London in the early 1860s as a representative of the American map and atlas publisher J. H. Colton. Because Bacon is known mostly for his pocket atlases and street maps of London, the world map projection for which he sought a US patent in 1912 (Fig. 5.11, bottom) probably reflects his sideline as a publisher of educational materials. Bacon was in his early 80s at the time, but still active in the business.[83] He described his projection as an attempt to counter the exaggeration of area at higher latitudes on the simple cylindrical and Mercator projections.[84] By contrast, his improved graticule remained rectangular below 70° latitude while curved meridians converge toward poles represented by points; another version used straight-line meridians near the poles. Perhaps because Bacon had already patented the map in Britain, his American application met little resistance, aside from the patent examiner calling his nine claims "unnecessarily numerous." Six were "rejected as involving merely an aggregation—an obvious one—of maps constructed according to two well-known systems of projection."[85] Although a shortened version was promptly approved, I have yet to find evidence that it was ever used.

Bacon's uncomplicated patenting of a largely obscure map projection contrasts strikingly with the success of Alphons Van der Grinten, a German-born Chicago draftsman who met marked resistance at the Patent Office but achieved prominence posthumously after the National Geographic Society cast its world reference maps on his distinctive projection. The front page of his only American patent (Fig. 5.12) integrates a circular world map (above) with a diagram (below) describing the construction of its meridians and parallels as arcs of circles.[86] Obvious exceptions are the prime meridian and the equator, plotted as straight lines. Meridians are evenly spaced along the equator, which has no distortion, but north–south scale increases toward the poles. The separation of consecutive parallels increases poleward, as does distortion of area—Greenland is almost as large as South America, as on the Mercator projection, and omission of Antarctica downplays the otherwise obvious exaggeration of polar area. J. Paul Goode praised Van der Grinten for "strik[ing] a happy mean between the extremes of the Mercator and the Mollweide"—projections that respectively preserved angles and area at the expense of rampant areal and angular distortion.[87]

Van der Grinten's patent application, submitted in October 1899, when he was 45 years old, was flagrantly flawed. Within a year after filing,

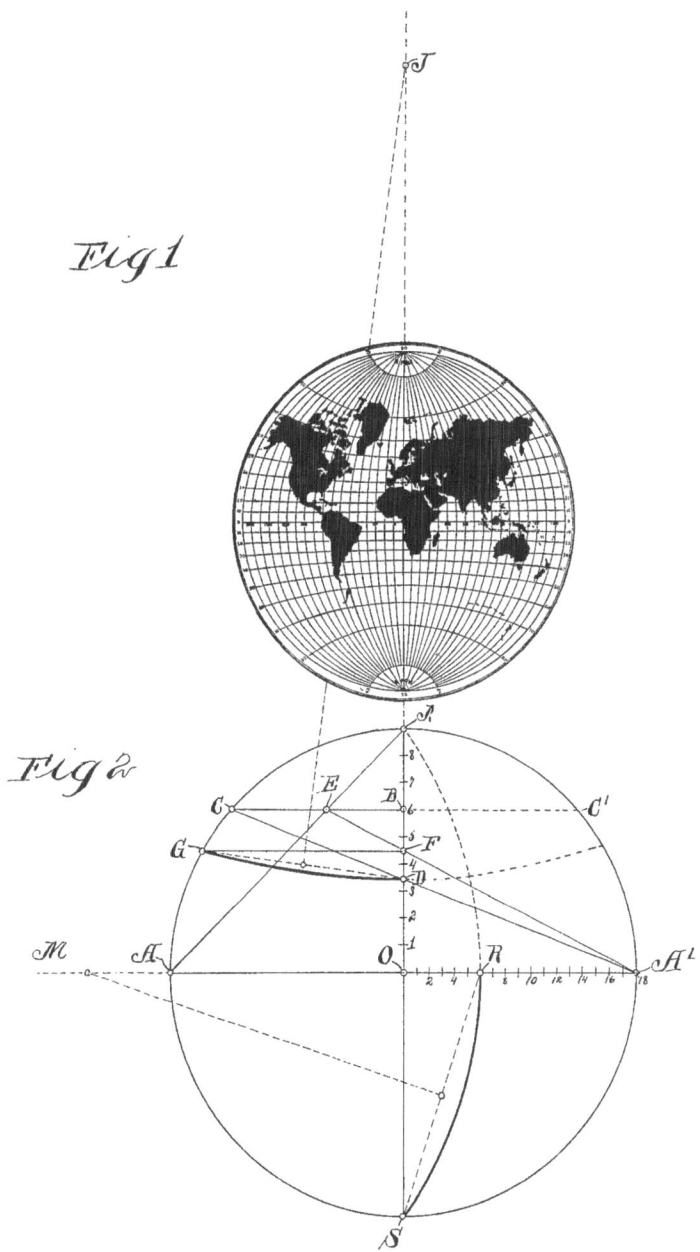

Fig. 5.12 Drawings in Alphons Van der Grinten's patent describe a worldwide map constructed from circular arcs for both meridians and parallels (US Patent 751,226; 1904)

the patent examiner had sent four letters of rejection, citing poor word-ing, lack of clarity, failure to describe a method (rather than the resulting map), and errors of fact like stating that the Mercator projection "was introduced in 1869," rather than 1569.[88] A lapse in correspondence of more than a year led the Patent Office to assume the application had been abandoned. In asking to have his case reinstated, Van der Grinten signed an affidavit blaming the lapse on poor health, lost income, and the need to work a six-day week at Rand, McNally & Co., his employer for "about twenty years," to support his wife and child.[89]

> That the duties at my regular employment with such firm are very tedious and arduous, and at the close of each day's work for the past two years and more, I have been so exhausted that it was necessary for me to remain in a state of absolute quiet and rest when arriving at home in the evening. That since April 2nd, 1901, I have been very feeble, and that I have been so tired and exhausted from following my regular vocation, that I was unable to per-form any services on my invention and improvements in maps ... and that the only time when I was in a physical and mental state to work upon such invention and improvements in maps was a few hours on Sunday morning, and that since April 2nd, 1901, I have constantly worked on Sunday morn-ings, when I was physically and mentally able to do so, upon such invention and improvements in maps, up to October 15th, 1902, and that I believe that through my persistent efforts, I have at last set out my invention and improvements in maps in such form that the Honorable Commissioner of Patents will readily see the merits of my invention.

The letter was attached to a formal request to reinstate, which was approved six weeks later.[90] In the meantime amended claims were sub-mitted, rejected, amended again, and ultimately approved in mid-July 1903.[91] Further correspondence included a request (promptly approved) that the original artwork be returned because "the analytical development [of the projection] is of such character as to render its production exceed-ingly laborious and the applicant has need of the same for the purpose of publishing it for the use of interesting publishers and users of maps in the future exploitation of the invention."[92]

Van der Grinten had no misgivings about the conventional scientific-technical literature, and promptly published two papers, one in Germany and the other in the *American Journal of Science*, in which he noted that the projection had been patented in Canada, France, and Great Britain; pro-vided both a "Continental View" (centered on 0°) and an "American View" (centered on 90°W); and presented formulas for calculating deformation.[93] He also devised three additional versions, which impressed John Snyder, who included images of all four projections in *Flattening the Earth*.[94]

None of the later versions attained the prominence of the Van der Grinten I, which won the approval of *National Geographic* editor Gilbert Grosvenor, who had asked his chief cartographer to recommend a projection with the low distortion of Goode's interrupted equal-area map and the familiar integrity of Mercator's uninterrupted world map.[95] For graphic efficiency, Grosvenor's cartographer severed the polar areas, thereby suppressing the projection's world-in-a-circle origin, but inserted small pole-centered maps in the upper corners. The map debuted in the December 1922 issue as a folded wall-map insert. A short accompanying article did not identify the inventor by name but claimed, "The World Map is the product of several years of research and labor."[96] The title block at the bottom of the map named the projection (but not Alphons) in tiny type and included the number of the patent, which had already expired—between 1861 and 1994 a US patent was valid for no more than 17 years.[97] Van der Grinten had died the previous year, at age 69, but even though his projection earned little if any income, it framed the Society's official world map for the next 66 years.[98] Snyder reported other adoptions, including economic maps published by the US Department of Agriculture and world maps in various textbooks.[99]

None of the three remaining patented projections merited mention in *Flattening the Earth*, but the USGS bibliography listed a patent titled merely "Map" awarded in 1926 to Samuel W. Balch, a mechanical engineer and patent attorney in Montclair, New Jersey, about 25 kilometers (15 miles) west of New York City.[100] Balch claimed to having devised a cartographic method for plotting a shortest-distance route between two points as well as for determining both the bearing and the distance from the origin at any point along the route. Departing from the oversimplified notion of the earth as a perfect sphere, he based his projection on an ellipsoid of revolution (a sphere flattened at the poles), which had become the basis for detailed topographic mapping, and offered two embodiments: an oblique Mercator projection and an oblique gnomonic projection. Ignored by the academic-scientific-technical literature, it was cited only twice in the patents literature—in the 1990s by examiners reviewing a pair of electronic mapping patents.[101]

Balch was 62 years old in 1924, when he applied for what would be the last of his 15 patents, most concerned with measuring time. Two years earlier, he had filed an application for a related invention titled "Ship's Course and Position Indicator," which was awarded a patent the same day as his map projection. I found no evidence that either invention was ever developed commercially.[102]

In what seems little more than a curious coincidence, Montclair was also home to William C. Anderson, who in 1936, 12 years after Balch, filed a tediously worded patent application titled "Map" for a projection intended as a navigation aid.[103] Anderson was 45 years old at the time and employed as a construction superintendent.[104] With four years of college, he had sufficient mathematical savvy to create a map for plotting a great-circle route using a protractor and the principles of conic sections. Two years later he sought patent protection for a more fully developed invention, appropriately titled "Method of Plotting Great Circle Courses and Apparatus."[105] While a set of templates made the method look workable, the patent was cited only once, by an examiner vetting an X-ray imaging method. Like Balch, Anderson gained little if anything from his second, closely related patent.

The last patent in my dataset is a method of map projection intended for depicting cities, rather than countries, continents, or the entire world. Earth curvature is far less troublesome for urban maps, which must cope with the typical downtown concentration of commercial, historical, and entertainment attractions. The problem is sheet size, which can become cumbersomely large when a street-map publisher seeks a legible and seamless treatment of the city and its surroundings on a uniform scale. Although the typical solution is a larger-scale inset focusing on interesting features near the city center, German map publisher Albrecht Falk supplemented his ingenious system of map folding with a map projection affording greater detail where it was needed most.

Twenty-six years old when he filed, Falk was the youngest of the inventors discussed in this chapter. His strategy was photographic and mechanical, rather than mathematical. Digital mapping had yet to enter the cartographic lexicon in the late 1940s, when he proposed a system of cleverly concocted distortions based on a wide-angle lens, multiple prisms, and a copy stand on which the base map bends away from the camera on a cylinder or a plane. Eager to prevent competitors from exploiting his ideas, he filed a patent application in early 1949 for a "Photographic Method for Making Geographical Maps."[106] His 16 drawings included an undistorted "lattice network" with uniform square cells (Fig. 5.13, upper left) and a lattice distorted by successive photographs, 90° apart on

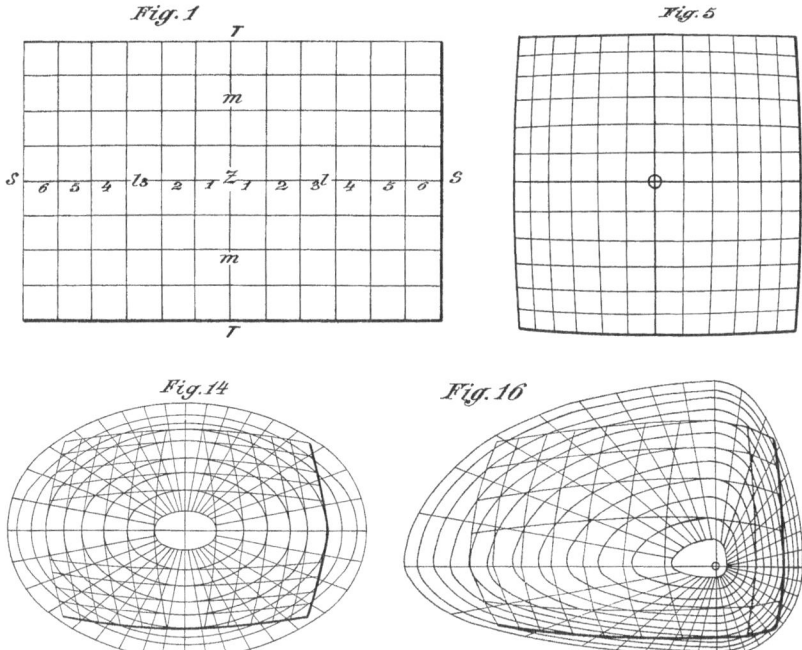

Fig. 5.13 Selected drawings from Gerhard Falk's patent of a photographic system for increasing a map's scale around the city center (US Patent 2,650,517; 1953)

a cylindrical platform, to provide a larger scale at the center (Fig. 5.13, upper right). Other drawings showed projected grids in which horizontal scale was greater than vertical scale (Fig. 5.13, lower right) and the area of greater detail was offset from the center (Fig. 5.13, lower right). Although textbooks and university courses on map projection might not consider Falk's clever geometric transformations a system of map projection, they were a purposeful and patentable invention.

It's clear that no single factor can account for the creation and patenting of a map projection: not mathematical expertise; not work experience as a draftsman, map publisher, or professional geographer; and not prior experience with the patent system—indeed, only five of the inventors had noteworthy prior patents (Balch, Boorman, Fisher, Fuller, and Spilhaus). But with the possible exception of Seattle IRS agent James Addison Smith, at least one of these factors seems relevant for the 16 inventors for whom microdata research tools yielded basic details about their lives and other accomplishments.[107] Among this group, the only discernible generalization is a vague relationship with age at time of filing: except for 26-year-old Gerhard Falk, all inventors were older than 40, with three (Balch, de Beaumont, and McBryde) in their 60s, two (Fisher and Spilhaus) in their 70s, and one (Bacon) in his 80s. These older inventors, if not conditioned to filing patents, no doubt considered intellectual property a desirable achievement. Moreover, none of the six (Cahill, Falk, Fisher, Fuller, Spilhaus, and Van der Grinten) who tried to profit from their patent seems to have made much money—further confirmation of John Snyder's insight.[108]

Notes

1. For a concise biography, see John W. Hessler, "Snyder, John P(arr)," in *Cartography in the Twentieth Century* (Vol. 6 of *The History of Cartography*), ed. Mark Monmonier, 1397–1400 (Chicago: University of Chicago Press, 2015). Map projections and the history of cartography had been Snyder's hobbies before he retired as a chemical engineer in 1980 and accepted a full-time research scientist position at the US Geological Survey. Although personnel records indicate he officially retired from the USGS in 1988, he continued to write and advise until his death in 1997. The quotation is from John P. Snyder, *Flattening the Earth: Two Thousand Years of Map Projections* (Chicago: University of Chicago Press, 1993), 302.

2. John P. Snyder and Harry Steward, *Bibliography of Map Projections*, US Geological Survey Bulletin 1856 (Reston, VA, 1988). The 14 patents are all US patents, but two were patented elsewhere. The *Bibliography* also lists a Russian patent published in 1968 by A. V. Borodin (entry 280).

3. Not covered by one of these categories is Bernard Cahill's second map projection (US Patent 1,081,207), shown in Fig. 5.3 and assigned to "Education and demonstration/Geography— Terrestrial globe or accessory therefor, collapsible or arranged for convenient assembly, disassembly, or storage" (434/137).

4. For the context of Stewart's approach to detecting hard-core pornography, see his short concurring opinion in *Jacobellis v. Ohio*, 378 U.S. 184, 197 (1964). The phrase seems to have been sug-

gested by Stewart's law clerk; see Peter Lattman, "The Origins of Justice Stewart's 'I Know It When I See It'," *Wall Street Journal*, 27 September 2007.

5. Snyder, *Flattening*, 143.

6. Ibid.

7. William Burke (examiner) to J. Marcus Boorman (c/o James A. Whitney), 19 August 1876.

8. James A. Whitney to Commissioner of Patents, 22 August 1876.

9. Burke to Boorman (c/o Whitney), 9 September 1876.

10. Whitney to Commissioner of Patents, 26 September 1876.

11. J. Marcus Boorman, "Improvement in Geometric Blocks for Mapping," US Patent 185,889, filed 26 July 1876, and issued 2 January 1877. According to the case file in the National Archives, the original title of Boorman's invention was "Improvements in Geometrical Solids for Mapping and Other Purposes."

12. David C. McClelland, *The Achieving Society* (Princeton, NJ: D. Van Nostrand Company, 1961), esp. 1–62, 225–28. Although McClelland's primary concern was economic growth, I doubt that I am drawing more meaning from his book than he had intended.

13. In the late nineteenth century, many, if not most, lawyers lacked a law school degree. Moreover, most lacked a college degree and some even lacked a high school diploma. See Judy G. Russell, "Tracing Legal Education 19th Century Style," *The Legal Genealogist*, 23 May 2012, http://www.legalgenealogist.com/blog/2012/05/23/tracing-legal-education-19th-century-style/. Also see Brian J. Moline, "Early American Legal Education," *Washburn Law Journal* 42 (2003): 775–802; and Robert Bocking Stevens, *Law School: Legal Education in America from the 1850s to the 1980s* (Chapel Hill, NC: University of North Carolina Press, 1983), esp. 3–13.

14. In 1887 *Mathematical Magazine* published collected lists of contributors and contents for Volume 1, released as 12 issues between January 1882 and October 1884. In addition to Boorman's occupation, the contributors list (on page iii) identified him as the author of notes, problems, or solutions that began on pages 84, 101, 112, 128, 169, 187, 204, and 207. An obsession with detail is apparent in this last contribution, titled "Square Root Notes," which began "THE value of [the square root of] 3 published in the MAGAZINE, pp. 165 and 172, is erroneous after the 174th decimal place."

15. J. Marcus Boorman, "Improvement in Hay-Rakers and Loaders," US Patent 77,247, issued 28 April 1868; "Improvement in Rotary Steam-Engines," US Patent 87,023, issued 16 February 1869; and "Improvement in Hay-Loaders," US Patent 101,705, issued 12 April 1870.

16. "Recent Patents," *Engineering: An Illustrated Weekly Journal* 7 (12 March 1869): 176. Boorman had filed for a British patent through London patent agent William Edward Newton. The full comment, "… a rotary engine of frightful and wonderful appearance, which we shall not attempt to describe here," was an editorial opinion, not wording supplied by Newton or quoted from the patent.

17. John Marcus Boorman, "Sailing-Vessel," US Patent 517,155, filed 1 August 1893, and issued 27 March 1894.

18. According to *Trow's New York City Directory* for 1880 and other years, Boorman's and Whitney's law offices were at 202 and 212 Broadway, respectively, in Manhattan.

19. William Wilson, "Geographical Map, Globe, and Other Geographical Appliance," US Patent 944,248, filed 28 July 1909, and issued 21 December 1909.

20. William Wilson, "Improvements in Geographical Maps, Globes, and Other Geographical Appliances," Great Britain Patent 190804419, filed 27 February 1908, and issued 29 March 1909. The council of the Royal Scottish Geographical Society acknowledged Wilson's donation in the minutes for its 3–4 October 1939 meeting, published in *Scottish Geographical Magazine* 55 (1939): 340.

21. Bernard J. S. Cahill, "Map of the World," US Patent 1,054,276, filed 5 March 1912, and issued 25 February 1913.

22. Cahill's respect for the integrity of continents was not shared by Indianapolis high school principal Walter Gingery, who patented a world map projection with North America inside a small circular map surrounded by six lobes resembling the petals of a flower. South America and Australia are largely intact but not Africa and Asia. Distinctive but hardly useful, the projection was listed in Snyder and Steward but not mentioned in *Flattening the Earth*. See Walter G. Gingery, "Map Projection," US Patent 2,352,380, filed 27 November 1942, and issued 27 June 1944.

23. See, for example, John Bartholomew, *British Imperial Federation Map of the World Showing Present Political Status of the Various*

Possessions of the British Empire, March 1889, Ca. 1:25,000,000 (Edinburgh: John Bartholomew & Co.; Edinburgh Geographical Institute, 1889). Online at http://nla.gov.au/nla.map-rm1392.

24. The patent describes the northern portions of the four lobes as "maintained in continuity along a parallel of north latitude 41 degrees, 48 minutes, 21 seconds [so that] if adjacent lobes [are] in contact with each other on said latitude, they will, when so distorted into a plane surface occupy an angular space of 240 degrees about the north pole. This permits the remaining angular space to be exactly filled by two other equal and similar ... lobes ... so that the map assumes a regular hexagonal form."

25. Bernard J. S. Cahill, "An Account of a New Land Map of the World," *Scottish Geographical Magazine* 25 (1909): 449–69; quotations on 455–56.

26. Bernard J. S. Cahill to Commissioner of Patents, 29 February 1912.

27. B. J. S. Cahill, "A World Map to End World Maps," *Geografiska Annaler* 16 (1934): 97–108; and B. J. S. Cahill, "Progress of the Butterfly Map—The Three Variants," *Architect and Engineer* 137 (May 1939): 47–49.

28. L. S. Underwood (examiner) to B. J. S. Cahill (c/o Francis M. Wright), 20 April 1912.

29. F. M. Wright to Commissioner of Patents, 4 June 1912.

30. Commissioner of Patents to B. J. S. Cahill (c/o F. M. Wright), 24 July 1912. The patent was not issued and published until early the following year, after the final fee had been paid.

31. Bernard J. S. Cahill, "Geographical Globe," US Patent 1,081,207, filed 11 February 1913, and issued 9 December 1913.

32. Snyder, *Flattening the Earth*, 264–65; quotation on 265.

33. For an example, see B. J. S. Cahill, "Projections for World Maps," *Monthly Weather Review* 57 (1929): 128–33, esp. 131.

34. "Cahill, B(ernard) J. S.," in *Who's Who on the Pacific Coast: a Biographical Compilation of Notable Living Contemporaries West of the Rocky Mountains*, ed. Franklin Harper, 89 (Los Angeles: Harper Publishing Co., 1913).

35. B. J. S. Cahill, "Projections for World Maps," *Monthly Weather Review* 57 (1929): 128–33; B. J. S. Cahill, "One Base Map in Place of Five," *Monthly Weather Review* 68 (1940): 41; and B. J.

S. Cahill, "Progress of the Butterfly Map—The Three Variants," *Architect and Engineer* 137 (May 1939): 47–49.

36. Nick Stockton, "Get to Know a Projection: Gene Keyes' 40-Year Quest for the Perfect Map," *Wired*, 9 December 2013, http://www.wired.com/2013/12/gene-keyes-quest-for-the-perfect-map/.

37. Gene Keyes, B. J. S. Cahill Online Resource, www.genekeyes.com/B.J.S._CAHILL_RESOURCE.html.

38. James Addison Smith, "Globe," US Patent 2,153,053, filed 22 November 1937, and issued 4 April 1939.

39. Snyder, *Flattening the Earth*, 270.

40. Joel E. Crouch, "Icosahedral Map," US Patent 2,424,601, filed 25 January 1944, and issued 29 July 1947.

41. Email communication from Meredith Anne Weber, Special Collections Library, The Pennsylvania State University, 6 November 2015. The Pennsylvania State College was granted university status in 1953.

42. Joel E. Crouch, "Astronomical Device," US Patent 2,412,130, filed 25 January 1944, and issued 3 December 1946.

43. Irving Fisher, "Global Map," US Patent 2,436,860, filed 19 February 1945, and issued 2 March 1948.

44. Irving Fisher, "A World Map on a Regular Icosahedron by Gnomonic Projection," *Geographical Review* 33 (1943): 605–19.

45. Irving Norton Fisher, the inventor's son, compiled *A Bibliography of the Writings of Irving Fisher* (New Haven, CN: Yale University Press, 1961), which lists two efforts to promote the projection. Item 2356 (p. 415), copyrighted in 1943 and "sold in 4 forms," is described as "Likaglobe, an Icosahedral world map, with instructions for folding into a near globe." Item 2373 (p. 419) is the reference for Irving Fisher, "A New World Map – Globe," *Click* 7.5 (May 1944): 27–29.

46. Irving Norton Fisher, *My Father Irving Fisher* (New York: Comet Press, 1956), esp. 160–61, 220–21. Several patents related to the card-index system were awarded in 1912. See, for example, Irving Fisher, "Index or File," US Patent 1,048,058, filed 29 July 1911, and issued 24 December 1912.

47. "Chronological List of Principal Books by Irving Fisher" in Fisher, *My Father Irving Fisher*, 340–42; and Irving Fisher and Osborn Maitland Miller, *World Maps and Globes* (New York: Essential Books, 1944).

48. Robert Loring Allen, *Irving Fisher: A Biography* (Cambridge, MA: Blackwell, 1993), esp. 277, 283–85.

49. Miller, O. M., "Notes on Cylindrical World Map Projections," *Geographical Review* 32 (1942): 424–30.

50. M. K. Peck (examiner) to Irving Fisher (c/o Pineles & Greene), 3 August 1945.

51. Pineles & Greene to Commissioner of Patents, 1 February 1946.

52. Snyder, *Flattening the Earth*, 269–70.

53. Robert W. Marks, *The Dymaxion World of Buckminster Fuller* (New York: Reinhold, 1960), 140–46; quotation on 143.

54. Richard Buckminster Fuller, "Cartography," US Patent 2,393,676, filed 25 February 1944, and issued 29 January 1946. The Buckminster Fuller Institute posted a list of Fuller's twenty-six patents, awarded between 1927 and 1983. See the Buckminster Fuller Institute, "About Fuller: Patents," http://bfi.org/about-fuller/bibliography/patents. Strictly speaking, the sixth and seventh were not issued until March and June 1944.

55. "Life Presents R. Buckminster Fuller's Dymaxion World," *Life* 14.9 (1 March 1943): 41–55.

56. "World Energy: A Map by R. Buckminster Fuller," *Fortune* 21.2 (February 1940): 57. The map was "executed by Philip Ragan," apparently to Fuller's specifications.

57. Strict standards existed as late as the 1990s, according to Richard Walker, who wrote, "Photographs or photomicrographs are acceptable but are restricted to illustrations of crystalline structures, metallurgical microstructures, textile fabrics, grain structures, and ornamental effects." See Richard D. Walker, *Patents as Scientific and Technical Literature* (Metuchen, NJ: Scarecrow Press, 1995), 142. Could a map or globe be an ornamental effect? A special pleading, not noted in the case file, might have occurred before the application was submitted insofar as both Fuller and his attorney were located in Washington, DC. The attorney's response to the first letter of rejection mentions an "oral interview [on 10 May 1944] courteously granted to [Fuller] and his attorney by the Examiner in charge of this application." Donald W. Robertson to Commissioner of Patents, 11 October 1944.

58. M. K. Peck (examiner) to Richard Buckminster Fuller (c/o Donald W. Robertson), 20 April 1944; F. J. Pisarra to Commissioner of Patents, 13 June 1945. Pisarra, an associate attorney at the firm,

had replaced Robertson as Fuller's attorney of record. Robertson to Commissioner of Patents, 14 February 1945.

59. M. K. Peck (examiner) to Richard Buckminster Fuller (c/o Donald W. Robertson), 14 December 1944.

60. Marks, *Dymaxion World*, 143–45.

61. Robert W. Gray, "Fuller's Dymaxion™ Map," *Cartography and Geographic Information Systems* 21 (1994): 243–46.

62. See Paul Knox and Sallie Marston, *Places and Regions in Global Context: Human Geography* (Upper Saddle River, NJ: Prentice-Hall, 1998), esp. 29–31, 121; and Mark Monmonier, *Rhumb Lines and Map Wars: A Social History of the Mercator Projection* (Chicago: University of Chicago Press, 2004), 170–71.

63. Jean Thorel, "Polygnomonic Map of the World Comprising Two Hemispheres," US Patent 3,868,781, filed 11 April 1974, and issued 4 March 1975.

64. Su Hui Wang, "Self-Assemble Revolving Globe," US Patent 4,620,842, filed 16 April 1985, and issued 4 November 1986.

65. Henri M. Dufour, "Map with a Homogeneous Grid System," US Patent 4,773,861, filed 16 March 1987, and issued 27 September 1988.

66. F. Webster McBryde, "Homolinear Composite Equal-Area World Projections," US Patent 4,315,747, filed 29 June 1977, and issued 16 February 1982.

67. E. Willard Miller, "In Memoriam: F. Webster McBryde, 1908–1995," *Annals of the Association of American Geographers* 86 (1996): 343–46; and Frank Canters, *Small-scale Map Projection Design* (London: Taylor and Francis, 2002), 65–66.

68. Snyder, *Flattening the Earth*, 218–19.

69. For examples see John P. Snyder and Philip M. Voxland, *An Album of Map Projections*, US Geological Survey Professional Paper 1453 (Reston, VA, 1988), 52–53; and Daan Strebe. "Mapthematics and Geocart Projections List – McBryde S3," https://www.mapthematics.com/ProjectionsList.php?Projection=62#McBryde S3.

70. William A. Nierenberg, "Athelstan Spilhaus, 25 November 1911· 30 March 1998," *Proceedings of the American Philosophical Society* 144 (2000): 343–47.

71. Athelstan Spilhaus, "World Ocean Maps: The Proper Places to Interrupt," *Proceedings of the American Philosophical Society* 127 (1983): 50–60; quotation on 52.

72. Athelstan Spilhaus and John P. Snyder, "World Maps with Natural Boundaries," *Cartography and Geographic Information Systems* 18 (1991): 246–54.
73. Athelstan Spilhaus, "Maps of the Whole World Ocean," *Geographical Review* 32 (1942): 431–35.
74. Athelstan Spilhaus, "Map Puzzle Having Periodic Tesselated Structure," US Patent 4,627,622, filed 15 March 1985, and issued 6 December 1986.
75. For photographs of Spilhaus's jigsaw puzzles and three-dimensional self-assembly polyhedrons, see Athelstan Spilhaus, "Plate Tectonics in Geoforms and Jigsaws," *Proceedings of the American Philosophical Society* 128 (1984): 257–69. A WorldCat.org search for "GeoLearning Corporation" returned a list of ten different games or visual aids, five of which are directly or indirectly linked to Spilhaus.
76. Advanced search of Google Patents with the name Athelstan Spilhaus in the inventor field.
77. More enigmatic is the attempt to define *periodic* by noting "a duplicate projection nests with the first projection to form a larger piece that itself is tessellated"—would *hierarchical* be a better word?
78. The three projections with excerpts in Fig. 5.11 are entries 520, 449, and 116 in *Bibliography of Map Projections* (1988).
79. Henry Bouthillier de Beaumont, "Map," US Patent 400,642, filed 15 October 1888, and issued 2 April 1889. For an obituary see Arthur de Claparède, "Henry Bouthillier de Beaumont, Président honoraire, Fondateur de la Société de géographie de Genève (1819–1898), Notice nécrologique" *Le Globe: Revue genevoise de géographie* 37.1 (1898): 1–14.
80. Jules A. Colas, "Map," US Patent 752,957, filed 24 February 1902, and issued 23 February 1904.
81. Colas also created a map of the moon (1894) and edited a celestial handbook (1893), both published by Poole Brothers.
82. David Smith, "George Washington Bacon 1862–c. 1900," *The Map Collector* no. 65 (Winter 1993): 10–15; and Ralph Hyde, "G. W. Bacon and His Atlases of London," introduction to *The A to Z of Victorian London* (Lympne Castle, Kent, UK: Harry Margary, 1987), v–viii; this book is a facsimile reprint of George W. Bacon, *New Large-scale Ordnance Atlas of London & Suburbs with Supplementary Maps, Copious Letterpress Descriptions, and Alphabetical Indexes* (London, 1888).

83. In his mid-30s Bacon had received an earlier patent, related to map printing. See George Washington Bacon, "Improvement in Constructing Blocks or Plates for Printing Maps," US Patent 57,056, issued 7 August 1866.
84. George Washington Bacon, "Map, Chart, and Geographical Diagram," US Patent 1,050,596, filed 4 May 1912, and issued 14 January 1913.
85. L. D. Underwood (examiner) to George Washington Bacon (c/o Munn & Co.).
86. Alphons Van der Grinten, "Map," US Patent 751,226, filed 2 October 1899, and issued 2 February 1904.
87. J. Paul Goode, "A New Method of Representing the Earth's Surface," *Journal of Geography* 4 (1905): 369–73.
88. Rejections sent to Van der Grinten's attorneys, Poole & Brown, in Washington, DC, were dated 11 October 1899, 8 May 1900, 18 May 1900, and 15 June 1900.
89. A notarized letter signed by Van der Grinten on 4 November 1902 accompanied a petition that Van der Grinten's attorney submitted to the Commissioner of Patents three months later.
90. Poole & Brown to Commissioner of Patents, 9 February 1903; and Commissioner of Patents to Van der Grinten (c/o Poole & Brown), 20 March 1903.
91. Van der Grinten to Commissioner of Patents, 9 February 1903 and 25 April 1903; A. P. Shaw (examiner) to Van der Grinten (c/o Poole & Brown), 5 May 1903; and Van der Grinten to Commissioner of Patents, 25 May 1903. Memorandum of final fee paid, dated 11 January 1904, reported that a Circular of Allowance had been issued on 15 July 1903.
92. Brown & Poole to Commissioner of Patents, 31 July 1903.
93. Alphons J. Van der Grinten, "New Circular Projection of the Whole Earth's Surface," *American Journal of Science* 19 (1905): 357–66.
94. Snyder, *Flattening the Earth*, 258–62.
95. Susan Schulten, *The Geographical Imagination in America, 1880–1950* (Chicago: University of Chicago Press, 2001), 195.
96. "The Society's New Map of the World," *National Geographic Magazine* 42 (1922): 690–91; quotation on 690.
97. Mark Lemley, "An Empirical Study of the Twenty-year Patent Term," *AIPLA Quarterly Journal* 22 (1994): 369–424. The term of patent increased from 17 to 20 years in 1995.

98. In 1988 the National Geographic Society replaced the Van der Grinten with the Robinson projection, replaced in turn in 1999 by the Winkel tripel projection. Monmonier, *Rhumb Lines*, 136–37.

99. Snyder, *Flattening the Earth*, 262. Snyder's principal source was Frank Kuen Chun Wong, "World Map Projections in the United States from 1940 to 1960," master's thesis, Syracuse University, 1965.

100. Samuel W. Balch, "Map," US Patent 1,610,413, filed 12 December 1924, and issued 14 December 1926. Balch's projection is entry 135 in *Bibliography of Map Projections* (1988).

101. See the citations list at the end of the online Google Patents pages for David M. Delorme, "Electronic Global Map Generating System," US Patent 4,972,319, filed 25 September 1987, and issued 20 November 1990; and C. Robert Pryor, "Map and Calculator Device," US Patent 5,902,113, filed 7 August 1997, and issued 11 May 1999.

102. Samuel W. Balch, "Ship's Course and Position Indicator," US Patent 1,610,412, filed 13 November 1922, and issued 14 December 1926.

103. William C. Anderson, "Map," US Patent 2,155,387, filed 22 August 1936, and issued 25 April 1939.

104. Anderson and Balch lived two miles apart. I found no evidence of any connection through a church, lodge, or other local organization.

105. William C. Anderson, "Method of Plotting Great Circle Courses and Apparatus," US Patent 2,268,632, filed 8 September 1938, and issued 6 January 1942.

106. Gerhard Ernst Albrecht Falk, "Photographic Method for Making Geographic Maps," US Patent 2,650,517, filed 10 February 1949, and issued 1 September 1953.

107. Minimal or better demographic details were found for 16 inventors: Anderson, Bacon, Balch, Boorman, Cahill, Colas, Crouch, de Beaumont, Falk, Fisher, Fuller, Gingery, McBryde, Smith, Spilhaus, and Van der Grinten. (Indianapolis high school principal Walter Gingery, whose his projection is mentioned here only in an endnote [note 22], was once a college mathematics instructor.) Inventors whose overseas residence or common surname thwarted searching are Dufour, Thorel, Wang, and Wilson.

108. Falk's patented map projection, though useful for street and tourist maps, was clearly less important to his business than his patented system of map folding.

CHAPTER 6

Global Affairs

Discussing world map projections before examining globes seems conceptually awkward insofar as scaling the coastlines on our three-dimensional Earth down to a manageable size is an essential preliminary to projecting them onto a flattenable surface like a cylinder, plane, or cone—at least that's what I tell my students. Treating map projections first might also seem historically awkward insofar as Gerard Mercator, the Flemish mapmaker who created the most famous and influential world map projection, was also a globe maker: a trade that helped him recognize that shifting the map's parallels progressively farther apart with increased distance from the equator could straighten out courses of constant direction called rhumb lines. But as Mercator and his contemporaries knew well, globe makers depended upon a supply of thin, tapering sectional maps called gores, each a low-distortion projection centered on a meridian. In a chicken-and-egg conundrum, the map projection comes first.

But unlike projections, globes are physical objects amenable to manufacturing, which makes patent protection useful to any inventor who wants to make and sell an innovative globe. It is hardly surprising, then, that the oldest US patent in my dataset for globes—number 2426, issued in January 1842 to Robert Piggot, of Elk Ridge Landing, Maryland, now a suburb of Baltimore—predates the first map projection patent (J. Marcus Boorman's geometrical solids) by nearly four decades. More significant, Piggot's patent followed by less than two years the first patent

© Mark Monmonier 2017

Mark Monmonier, *Patents and Cartographic Inventions*,
Palgrave Studies in the History of Science and Technology,
DOI 10.1007/978-3-319-51040-8_6

in my printed maps dataset, titled "Art of Hydrographic Surveying" and issued in June 1840 to H. Ariel Norris.[1]

Strangely, Piggot's patent includes no drawings, at least not in the scanned patent downloaded from Google Patents. A one-sentence note, attributed to Finis D. Morris, Chief of Division E, and dated 25 October 1913, preceded the printed text: "A careful search has been made this day for the original drawing or a photolithographic copy of the same, for the purpose of reproducing the said drawing to form a part of this book, but at this time nothing can be found from which a reproduction can be made."[2] Although whatever physical models Piggot provided were probably lost in one of the Patent Office's devastating fires, the artwork might have survived, later to be misfiled or swiped. Aware of this possibility, a helpful archivist examined folders for nearby patents but could not find it.[3]

Piggot's verbal description partly compensates for the missing artwork. His patent's title, "Apparatus for Teaching Geography and Astrography," establishes a pedagogic goal common to many globes. The first of its eight figures, he tells us, "is a representation of the kind of sphere which I employ both for geographical and astrographical delineations." (*Astrography* is an obsolete term for mapping the heavens.) He then describes a hollow globe, crafted from wood or paper, and divided into two equal parts, which can be joined together. On the outside are "grooved indentations … corresponding with the greater or lesser circles usually delineated on the terrestrial globe"—that is, the typical spherical grid, or graticule, of meridians and parallels. On the inside is another set of inscribed lines "ordinarily drawn on the celestial globe"—lines representing the constellations, named for mythical figures or objects like the Big Dipper, as well as the tropical and polar circles and the ecliptic, a circle that describes the sun's apparent path around the earth. Both sets of lines are indented below a surface of "artificial slate," to be drawn upon with "white crayons" (chalk) and "readily erased." The patent also covers an alternative light-colored surface on which lines drawn with "a lead pencil" can be "erased by india rubber." These indented lines comprise a framework for "the scholar [who] is required to draw … the outlines of the respective continents, islands, or other divisions of the globe, in their relative positions"—a hands-on approach to a (thankfully) bygone era of geographic education centered on rote memorization.

Piggot knew about incising lines. His patent describes him as a "minister of the Gospel," but before his ordination in the Protestant Episcopal Church in 1823, at age 28, he had worked in New York as an engraver, specializing in portraits.[4] He was also acclaimed for his handwriting and drawing ability, and although I found no evidence that he ever made maps or taught geography, in 1851, while living in Baltimore, he taught drawing at Western High School, the city's leading public secondary school for women.[5] In 1856 he was the professor responsible for graphics and fine arts at Newton University, a short-lived college in the city.[6] In 1869 he was called to a parish in Sykesville, about 25 miles west, and remained there until his death in 1887, at age 90. Although Piggot engraved for several decades after ordination, he apparently never manufactured his globe or licensed his patent.[7]

By contrast, Silas Cornell, who in 1845 received the second American patent for a globe, actively marketed "Cornell's Improved Terrestrial Globe" to educators in particular as well as to the general public.[8] And he had a broader reputation as a mapmaker, having worked as a surveyor and civil engineer in Rochester, New York, where he served as city surveyor between 1835 and 1839, and compiled and drafted detailed city maps, updated as the city grew, from 1845 to 1884.[9] Cornell was also an educator, having taught school on Long Island before moving to Rochester in 1823, at age 33, and running a private school there with his wife, also a teacher, on land outside the city that he cleared for a farm and nursery business. The idea for the globe probably reflects his efforts to teach basic geographic concepts like seasonal differences in solar radiation.

No less important than the globe was its 36-page instruction book, verbosely titled *A Description of Silas Cornell's Improved Terrestrial Globe with the Manner of Using It: Intended for the Use of Schools, Academies, and Families.*[10] In straightforward English, a marked contrast to the convoluted legalese of his patent, Cornell describes the key elements of the globe and its stand, packed separately to avoid damage, and discusses its use in explaining how earth-sun relationships affect the length of day and night at different times of the year at different places. Testimonials from four professors, a college president, and the Episcopal Bishop of Western New York precede the preface, in which Cornell emphasizes the globe's "new construction, differing materially from all others heretofore in use" and attributes its "great advantage in giving the first lessons in geography" to his "many years in teaching."[11]

A key feature is a thin metal disk called the "day-circle," which stands vertically, swivels about the base, and divides the globe into a "day side," where the sun is shining, and a "night-side," where it isn't. An attached pointer links the day-circle to a circular calendar, called the "index," on the

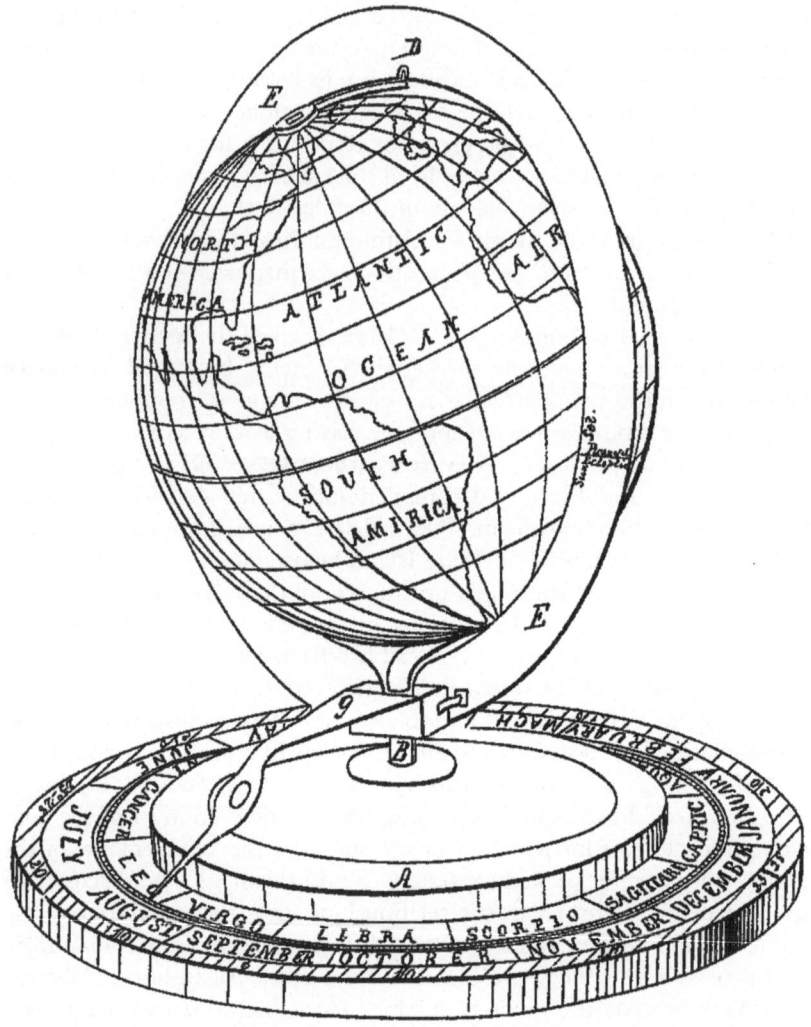

Fig. 6.1 Silas Cornell's globe used a movable "day-circle" (E) to relate the period of daylight to the month and day, inscribed on the circular "index" on the base of the mounting. The axis was fixed, to represent the fixed inclination of the earth's axis to the plane of its orbit about the sun. The globe can rotate on its axis independently of the day-circle and its pointer (g), which can rotate through 360° to describe the progression of seasons throughout the year (US Patent 4,098; 1845)

base of the frame. A drawing opposite the title page illustrates the alignment of the day-circle for mid-August, when places north of the equator enjoy more than 12 hours of daylight and latitudes near the North Pole have a full 24 hours of daylight. The corresponding drawing in the patent (Fig. 6.1) is identical, except for more delicate linework and letters marking specific features. The patent's title, "Mounting Globes," implies that the globe's frame, not the ball coated with graticule and coastlines, is the crux of Cornell's invention.

Evidence that Cornell actively marketed his invention include a small advertisement in the July 1844 issue of the *District School Journal of the State of New York* and a short review in the November 1846 issue of the *Common School Journal*, edited and published in Boston by the influential educator Horace Mann. The 1844 ad announced a "beautiful and cheap instrument, 5 inches in diameter, accompanied with a card of lessons" covering the topics in Cornell's 36-page book; the price was $1.50, with "a liberal discount allowed to dealers."[12] A downsized, less detailed version of the engraving in Fig. 6.1 shows the globe without its day-circle, pointer, and circular index—a low-cost, stripped-down version of the apparatus described in the patent. Perhaps Cornell believed a less expensive version could be sold to individuals as well as to schools: his ad not only argued that the globe's "cheapness renders it admissible to every school" but also claimed it belonged "in every school and every family." Two years later, a higher price point (three dollars), no doubt still affordable to schools, if not families, was endorsed by a reviewer identified as Galileo, who began, "This cheap little affair is really one of the happiest inventions that we have seen for many a day."[13] Impressed by its versatility as well as its price, the pseudonymous critic asserted, "it performs all the problems usually taught with the most expensive globes far better than they do, while it explains other phenomena, not taught by them, with beautiful simplicity."

I have no idea how many globes Cornell sold, but copies turned up decades later in several prominent collections, including the David Rumsey Map Collection, which posted a scanned image online. A catalog record describes "an uncolored 4.5″ globe on a wooden base … encircled [in turn] by a calendar in conjunction with the astrological signs" and reports it was produced in 1845 and "made and sold by Silas Cornell, Rochester, N.Y."[14] Like the small globe in the ad, it is missing the day-circle and pointer. By contrast, the Smithsonian Institution owns a larger version with an aluminum "meridian circle" and pointer, which resembles the patent drawing. According to the accompanying discussion, Cornell sold a

5-inch globe for $3.50 and a 9-inch globe for $10, and Erastus Darrow and Brother, a Rochester bookseller and publisher, was the actual manufacturer. Although Cornell had been marketing globes in the early 1840s, before he received his patent, the Smithsonian's globe was assembled after 1850. The gores were printed in black and colored by hand—a typical practice for mid-nineteenth-century atlases and wall maps. Isotherms (lines of equal temperature) that mark the northern and southern extent of wood production, grain, wine grapes, and bananas are the first appearance of "geophysical information on an American globe."[15]

Two other Cornell globes survived in collections canvassed by Ena Yonge for her *Catalog of Early Globes, Made Prior to 1850 and Conserved in the United States: A Preliminary Listing*.[16] Yonge was map curator at the American Geographical Society, in New York, from 1917 until her retirement in 1962.[17] Her book, published by the Society in 1968, was the American component of an international effort, begun in 1952, to inventory antique globes. She found a 5-inch version at the New York State Library, in Albany. Sadly, it was "in poor condition, the globe being apart from its base of wood." Another 5-inch globe, "in excellent condition," was in the good hands of Mrs. Orrin S. Thompson, of Red Hook, New York, midway between Albany and Manhattan. Yonge reported, "The globe revolves on a metal pedestal with a Maplewood base. A metal strip [the day-circle?] fits on the globe, marked off with figures of sunrise, etc."—if not currently in an institutional collection, it's a good candidate for Antiques Roadshow.

Before moving on to a systematic analysis of globe patents, I want to look at two other pedagogically inspired globes, both patented by women. Ellen Eliza Fitz, who was born in New Hampshire around 1835, left a more lasting impression. Her biography, pieced together largely from manuscript Census schedules, is surprisingly sketchy, given the cleverness of her invention and the thoroughness of her 120-page *Hand-Book of the Terrestrial Globe, or, Guide to Fitz's New Method of Mounting and Operating Globes, Designed for the Use of Families, Schools, and Academies*, published in 1876.[18] Despite its lengthy title, *Hand-Book* is more a textbook on globes and geography than a user's manual for Fitz's invention. Long out of copyright, it survives in reprint editions from over a half-dozen print-on-demand publishers.[19]

When she obtained her patent, in 1875, Fitz was working as a governess in St. John, New Brunswick, Canada.[20] I have yet to find any record of where she attended school, but a likely influence is Emma Willard, the nineteenth-century pedagogue and atlas author who promoted geographic education and mapmaking in the new nation's female academies.[21] Ellen might have acquired an early interest in globes and publishing from her father, Asa Fitz, who was an editor and textbook author.[22] Known mostly for compiling hymnals, he also co-authored *An Elementary Geography for Massachusetts Children*, published in 1845—a questionable inspiration insofar as its well-illustrated chapter "The Earth or Globe" does not go much beyond latitude, longitude, rotation, and revolution.[23] By contrast, Ellen's *Hand-Book* offers an impressively lucid description of earth-sun relationships and their effect on solar heating and the length of day and night.

Fitz's globe was understandably more sophisticated than Cornell's, patented three decades earlier. Each globe is mounted within a circle representing the calendar year, has its axis of rotation tipped 23½° away from the vertical, and treats the table on which it sits as the ecliptic plane, in which the earth moves around the sun. But, as Fig. 6.2 shows, in Fitz's invention horizontal pointer *e*, mounted on rod *d*, indicates the place at which the sun is directly overhead at noon. Two vertical "standards" (incomplete circles *H* and *I*) distinguish the regions of total darkness and twilight: the area to the right of *I* (similar to Cornell's day-circle) is in total night, while the area between *H* and *I* receives some light because of atmospheric refraction. In addition, two brass rings—ring *K* and semi-ring

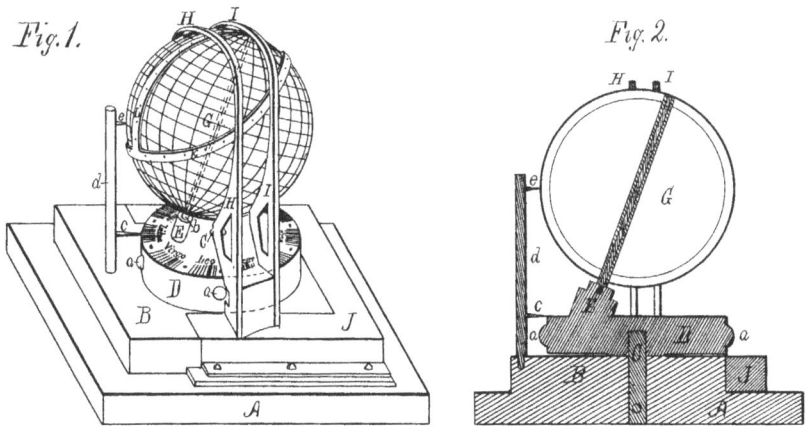

Fig. 6.2 Ellen Fitz's globe used two circular standards to differentiate areas in full daylight, twilight, and total darkness (US Patent RE 9,557; 1881)

L, called the *brass horizon* and *brass meridian*, respectively—can be, as Fitz explains in her book, positioned "to represent the horizon and meridian of any given place."[24]

To illustrate Fitz's globe, I had a choice of three somewhat different images. She had filed her original application in December 1874 and received a patent with remarkable dispatch in January 1875. After filing an amended application in May 1879, she received a so-called reissue patent in February 1881.[25] Reissue patents correct unintentional defects in language or artwork but do not extend the term of the patent—I chose these latter's drawings for Fig. 6.2 because of their larger, more legible labels.[26] Eager to claim further improvements, including a device showing the positions of the constellations and other stars at any time of the year, Fitz filed a new application in November 1881, and was rewarded with another patent, with a new 17-year term, in September 1882.[27] She died four years later, at age 51, in eastern Massachusetts, where she had moved in the late 1870s.[28]

The impact of Fitz's globe is confirmed by its preservation in diverse collections, including the Library of Congress, the Smithsonian Institution, and the Boston Public Library. Ginn and Company, the Boston firm that published the *Hand-Book*, sold 6- and 12-inch models. Even the smaller, less expensive version is rare and much sought after, as suggested by the sale of a 6-inch Fitz globe in October 2014 at the New York auction house Bonham's for $4375. Despite "some general light soiling to [the protective] varnish [and] a few repaired cracks and tears," it was "still a very attractive and unusual globe."[29] Selling points included its mounting, highlighted as a "special feature," and its prominence as "the first globe to be designed by a woman."

Which is not correct: other nineteenth-century women had made globes, and one even patented one, in 1831, more than four decades before Ellen Fitz. That inventor is Elizabeth Oram, who is mentioned by several authors, whose citations trace back to Deborah Jean Warner, a curator of scientific instruments at the Smithsonian.[30] On part of a single page of one of a series of six articles titled "The Geography of Heaven and Earth," Warner offers a tantalizing two-sentence summary:

> The author of several text books and books on pedagogy, Elizabeth Oram ran a 'Ladies Academy' in New York. In 1831 she obtained a patent for a 'globe for teaching geography'.[31]

Unfortunately, there is no citation of the patent, nor can there be: the 1836 fire that consumed all granted and pending patents as well as 9000 drawings and 7000 models left no record to be scanned for the Google Patents Project.[32]

Efforts to reconstruct the 10,000 incinerated patents led to what's called X-patents—patents restored from the owners' records and assigned to a new series with numbers starting at 1, each followed by the letter X.[33] The Directory of American Tool and Machinery Patents, a project run by volunteers interested in the history of woodworking tools and early American machinery, maintains an online list that includes patent 6,337X, titled "Globe for Teaching Geography" and granted to Elizabeth Oram on 12 January 1831. Unfortunately, the entry says, "Little is known about this patent. There are no patent drawings available. This patent is in the database for reference only."[34]

I found the same hazy evidence in *A Digest of Patents Issued by the United States, from 1790 to January 1, 1839*.[35] The only new information was its assignment to "Class XVIII—Arts, Polite, Fine, and Ornamental," a genteel category that included "Music, Painting, Sculpture, Engraving, Books, Paper, Printing, Binding, Jewellery [sic], etc." Globes and teaching aids were apparently part of the "etc." But HathiTrust.org, the rich online archive of arcana in which I found the *Digest*, also yielded an 1831 issue of the *Journal of the Franklin Institute*, the full title of which ends with … *the Recording of American and Other Patented Inventions*. Its list of "American Patents for January, with Remarks," reports "an Instrument for the Teaching of Geography; Elizabeth Oram, city of New York, January 12."[36]

Oram's entry begins with a mix of admiration and condescension: "As this instrument is the invention of a lady, we will, of course, allow her to tell her story in her own way, without any animadversions of ours, which mar the narrative, or involve us in inextricable difficulties." There's no drawing, of course, and Oram's original text, which follows, was probably longer and more detailed. Nonetheless, the clarity of her uncorrupted, unusually elegant pre-patentese is worth quoting.

> Be it known, that I, Elizabeth Oram, of the city of New York, have invented a new and useful instrument for the teaching of geography, and that the following is a full and exact description of the construction and use thereof as invented by me.
>
> It consists of a globe, upon which the surface of the earth is represented by various heights, as they exist in nature. By this the distinction between land and water is clearly seen. The various ranges of mountains, with their relative heights exhibited; and their influence upon heat and productions, with their geologic structure.

By means of a magnet inserted in the surface of the improved globe, the great principle of attraction may be clearly shown, by affixing thereto any small iron figure.

This globe is surrounded by the principal circles of the sphere. The ecliptic is elevated, by means of which, and a [*sic*] moveable and illuminated sun, the manner in which the earth receives its rays, and the causes of the seasons may be clearly exhibited. On the horizon there is affixed a small instrument, by which the causes of eclipses are shown.

A movable star, brings to the comprehension of pupils the nature of right ascension, declination, celestial latitude and longitude.

ELIZABETH ORAM.

Impressive but not easy to mass produce, especially in the 1830s, which probably explains why I found no record of Oram's globe in a collection or auction report. Despite the absence of an obituary or a biography, which might have documented her invention's success, however limited, archival evidence confirms that she was principal of New York's Female High School in 1830 and author, decades later, of *Oram's Grammar*.[37] Although her experience as an educator no doubt influenced her invention of a globe, her prominence was insufficient to bring the idea to market. I doubt that Oram's globe saw much use outside her own school.

Nonetheless, her invention, although apparently never developed, was a substantial achievement for the early 1830s. Moreover, because a text search of the *Digest of Patents*, covering the period 1790 through 1839, failed to find another instance of a patented terrestrial globe, Elizabeth Oram was not only the first woman to patent a globe but also the first American to patent one as well—11 years ahead of Robert Piggot. And because her second paragraph suggests that she had invented a relief globe—one of the 18 relatively specialized types of globes recognized by the US Patent Classification System—her invention probably predated the first patented relief globe by three decades.[38]

As the uneven and limited success of Oram, Piggot, Cornell, and Fitz implies, during the first half of the nineteenth century, globes were prominent mostly as classroom teaching aids, designed to mimic the precision of scientific instruments. This usage declined during the latter half of the century, as globe makers largely abandoned day-circles and intricate mountings and shifted their attention to less complicated varieties marketed as decorative objects for homes and offices. By 1900 globes had become as commonplace in middle-class homes as world atlases and encyclopedias, and also important as symbols of an enlightened interest in foreign affairs.[39] Jan Mokre, head of the Map Department and Globe

Museum at the Austrian National Library and author of the entries on globes in the twentieth century volume of the *History of Cartography*, was impressed by the large size—up to 106 cm (42 inches) in diameter—of globes manufactured in the 1930s, when they reached peak prominence as social and geopolitical symbols. Despite the continued persistence of globes in political and corporate iconography as well as the omnipresent and overworked adjective *global*, the importance of the three-dimensional globe declined markedly toward the end of the century, when the interactive, zoomable "virtual globe" eclipsed its non-electric counterparts.

This expansion and contraction is apparent in Fig. 6.3, which describes the number of US patents issued for terrestrial globes, by year, between 1831 and the second decade of the twenty-first century. Based on year of

Number of U.S. patents for terrestrial globes, by year of issue, 1831 to 2015

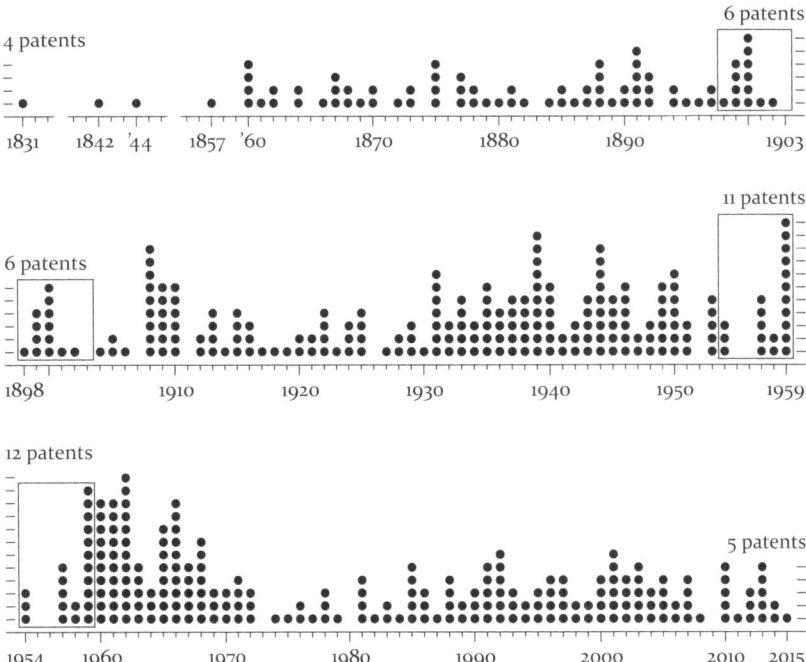

Fig. 6.3 Time-series plot of patents for terrestrial globes, by year of issue, 1831–2015. Each dot represents one patent in class/subclasses 434/131–148. Splitting the graph into three slightly overlapping parts promotes legibility, and thin identical rectangles identify areas of overlap. Compiled by author

issue—application dates were not available for the earliest patents—the graph is divided into three sections to avoid compressing the 186-year interval into an overly narrow, less informative display. (As with Fig. 1.4, thin rectangles marking areas of overlap are intended to avoid misinterpretation.) Aside from a comparative dearth of patents for globes throughout most of the nineteenth century and less prominent activity after the early 1970s, the two most potentially significant patterns are a 31-year period of moderate-to-high patenting between 1931 and 1951 and a shorter, more intense period between 1957 and 1972. Lack of a ready socio-political or pedagogic explanation makes these clusters no more significant than their bordering episodes of lower activity, and although the minor burst of patent grants following the Spanish-American War (1898) might reflect the new American presence in the Pacific Ocean, the causal connection is weak at best. Whatever explanations exist more likely reflect changes in the technology of globe making.

To better understand the uneven granting of patents for globes, I turned to the US Patent Classification System, which assigns terrestrial globes to 18 subclasses under class 434, Education and demonstration. As listed in Table 6.1, 17 of the 18 subclasses are indented below (and thus subordinate to) the more general and more common subclass "Terrestrial globe or accessory therefor" (with 125 patents), and 2 of these 17 (434/138 and 434/140) are further indented. Columns to the right of the category names show the total number of patents so assigned as well as the year of the earliest and most recent patents in the subclass and the median year. For example, of the 22 relief globes (subclass 434/132) in the dataset, the earliest patent was issued in 1861, the latest was issued in 2014, and the median year of issue is 1973—making relief globes perhaps the least temporally concentrated of any of the subclasses.

Technological influences are readily apparent for several of the subclasses. Although terrestrial globes "with means representing [a] vehicle moving relative to earth" (subclass 434/139) were patented as early as 1900, the indented category "Space vehicle" (subclass 434/140) did not see its first patent until 1958, the year after the Soviet Union launched Sputnik I. And the first patent in the category "With internal light" (subclass 434/145) was not issued until 1914, around the time most urban homes were connected to the power grid. By contrast, the more pedagogically relevant subclasses "Relief globe" (434/132), and globes "with means indicating time at different points on earth" (434/142) and "with means demonstrating solar illumination of earth" (434/143) saw their

Table 6.1 Size and date ranges for subclasses 434/131 through 434/148 within USPCS class Education and demonstration

Subclass		Size	Earliest	Median	Latest
131	Terrestrial globe or accessory therefor	125	1842	1947	2010
132	Relief globe	22	1861	1973	2014
133	Having diverse use (e.g., pencil box, etc.)	22	1860	1941	2005
134	Having magnet associated therewith	18	1864	1980	2015
135	Having plural planar or curved surfaces (e.g., flat or frustoconical surfaces)	26	1909	1978	2012
136	Rotated by mechanical drive	37	1873	1988	2013
137	Collapsible or arranged for convenient assembly, disassembly, or storage	38	1866	1967	2010
138	Inflatable	27	1868	1929	2012
139	With means representing vehicle moving relative to earth	12	1900	1940	1979
140	Space vehicle	37	1958	1962	1983
141	With means indicating distance between points on earth	17	1908	1963	2012
142	With means indicating time at different points on earth	50	1860	1935	2014
143	With means demonstrating solar illumination of earth	39	1860	1949	2003
144	With means demonstrating wind currents over earth	6	1886	1940	1991
145	With internal light	53	1914	1950	2005
146	With means to facilitate finding or reading indicia thereon	18	1888	1986	2005
147	With map segment attachable thereto (e.g., continent, nation, etc.)	29	1888	1991	2011
148	With suspension type support	15	1860	1910	2012

All are subordinate to subclass 130, Geography. Subclassess 132–148 are subordinate to subclass 131, and subclasses 138 and 140 are further indented. Because some patents were assigned to more than one subclass, the sum of the class sizes (591) exceeds the number of patents (475) in the dataset.

first patents in 1860 and 1861, when many inventors were still focused on classroom usage. Oddly, Silas Cornell's 1852 patent, even though its day-circle identifies areas receiving sunlight on any particular day, was assigned to the more common, least specific terrestrial globe subclass (434/131), perhaps because its drawings looked less sophisticated.

The first patent for an "Inflatable" globe (434/138), awarded in 1868 to Gorham D. Abbot, reflects a strange conflation of pedagogy and technology insofar as Abbot, a New York City clergyman, not only ran a collegiate institute for women but was also a major investor in the American Hard Rubber Company, which he had sued vigorously earlier in the decade.[40] Abbot's inflatable globe was far from a one-piece geographical balloon. His patent called for printing the globe "in lune-like sections, in the ordinary way, on any material combining flexibility with strength and durability" and then sewing or cementing them together. Although "a parchment or closely-woven fabric" might work "very well," he recommended "running a very thin tissue of rubber upon cotton, linen, silk, or mixed goods." I found no record that Abbot ever developed his invention, which seems prone to slow leaks. He died six years after receiving the patent. The latex balloon globes occasionally found in toy stores were decades away.

Abbot's inspiration might have been an inflatable globe included in the Great Exhibition at the Crystal Palace in Hyde Park, London, in 1851. A published description of the exhibition included a section on globes, in which a single enigmatic sentence reported, "Goodyear, of the United States, exhibited inflated globes two feet in diameter, of India-rubber or silk, varnished with the former material—also India-rubber maps."[41] Apparently, the first American inflatable globe preceded the first American patent for an inflatable globe by more than a decade and a half.

Although no first name was provided, the Goodyear mentioned was almost certainly Henry Bateman Goodyear, a brother of Charles Nelson Goodyear, who had invented vulcanized rubber around 1840. Henry worked in his brother's business and was no doubt acquainted with the patent system. He was also most certainly aware that hard rubber balls provided an alternative substrate for conventionally manufactured globes, produced by pasting printed globe gores onto a papier-mâché sphere. This realization along with his apparent interest in terrestrial globes probably inspired his own patent, awarded in 1861 for a markedly different invention. Titled "Method of Relieving Geographical Outlines on Molded Plastic Globes," it is the earliest patent in the Relief globe subclass (434/132).[42]

Henry's insight was to treat the rubber ball as the globe, not as a mere substrate. He understood how a perfectly spherical ball of India rubber (natural rubber) or gutta-percha, produced by fusing together two molded hemispheres, could first be "cut or reduce[d]" to make land stand out from water "so that not only is a child able to see and examine the several divisions and objects on its surface, clutching and feeling as well as seeing … the rotundity of our globe, but, whereby also, the poor blind child, may in a similar manner [learn] the same truths." While his notion of "relief" did not include the more intricate shapes of mountain ranges and plateaus, he recognized that a relief globe, like relief printing, could also include place names inscribed in raised type.

Although a sentence in *Knight's American Mechanical Dictionary*— "Globes are also made by Goodyear, of inflated india-rubber, or silk coated with india-rubber solution"—suggests that Goodyear might have manufactured and marketed an inflatable globe, I found no evidence that his relief globe was ever produced commercially.[43] At least a fruitless online search using "rubber" and "relief globe," or for similarly promising combinations like "Goodyear" and "globe," uncovered no advertisement, catalog entry, news report, or other trace of a retail offering, and no evidence that he had licensed the patent or had any effect on commercial globemaking. According to Jan Mokre, globe manufacturing in the late nineteenth and early twentieth centuries relied on hollow, spherical papiermâché substrates assembled from hemispheres crafted layer-by-layer on a wooden mold.[44] After 1910 hydraulic presses molded comparatively smooth hemispheres, and after 1930, American globe makers perfected a method for printing projected hemispherical maps onto cardboard disks, which were then cut into pinwheels and pressed into hemispheres under great heat and pressure.

Although production processes fall outside the realm of patented globes, a particularly intriguing invention is the hand-operated "Printing Machine for Spheres and the Like" invented in the late 1920s by Francis Augustus Lovegrove, of Halifax, Nova Scotia.[45] Born in England in 1869, he moved to Halifax in 1888. The 1921 Canadian census recorded his occupation as draftsman, which accounts for the clarity and precision of the eight drawings that illustrate his US patent, awarded in September 1929. His first drawing (Fig. 6.4, left) describes a hand-cranked machine that prints lines and labels onto a blank sphere from above and below. The image for the upper hemisphere is transferred from "concave leaves 34, the surface of each leaf being formed in spherical sections engraved or adapted to carry type" representing "an atlas [map] of the world." A corresponding set of concave leaves 44 at the bottom of the machine holds an image for the southern hemisphere. The operator inserts a blank sphere between the two sets of leaves, and rotates handle 21 to force the two sets of concave leaves together, onto the sphere, so that "the impression of the atlas will be made on the globe in either plan or raised type." His eighth drawing (Fig. 6.4, right) is a "top plan view" of the map formed on the globe's upper hemisphere.

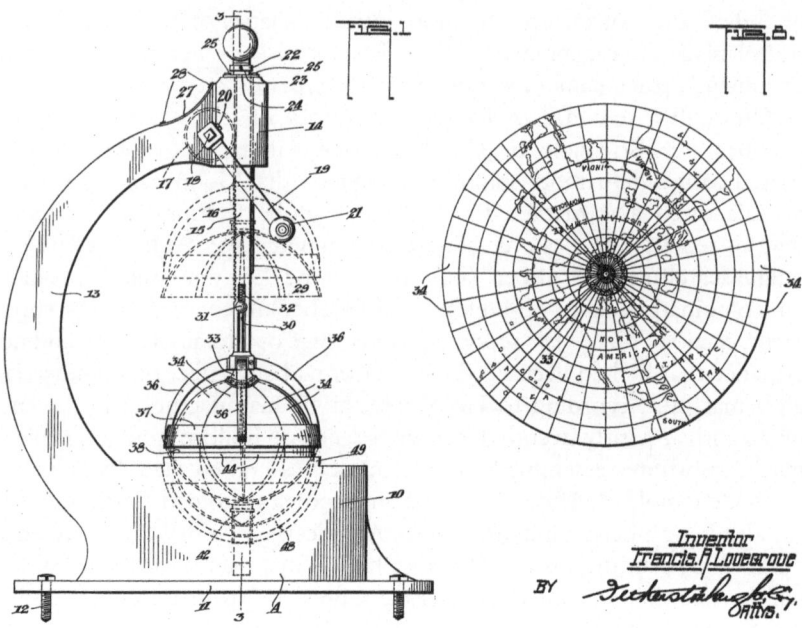

Fig. 6.4 Key elements of Francis Augustus Lovegrove's machine for printing a globe's graticule, coastlines, and labels onto a sphere (US Patent 1,728,351; 1929)

Lovegrove, who later became a government surveyor, was a sporadic inventor familiar with the patent system. Earlier inventions include a mundanely practical pipe threading and cutting machine, patented in 1917, and a Rube Goldberg-like apparatus, patented in 1914, whereby a stricken or otherwise submerged submarine can communicate with the surface through a tethered buoy with a telephone, a spring operated flag, a remotely controlled revolver that fired blanks to attract attention, and a flexible hose for emergency access to fresh air.[46]

Like many inventors, Lovegrove appreciated secondary uses. For example, although his globe-printing machine was intended to cut production time, lower cost, and minimize tedium, "another and essential object of the invention is to provide a globe atlas of the world in raised type to afford blind people the same facilities for learning geography as those who have sight." Unaware that Henry Goodrich had proclaimed a similar goal nearly seven decades earlier, Lovegrove believed he had discovered an approach "that has not heretofore been known or disclosed." While I

Table 6.2   Number of patents (n), by subclass, in each of the two notable time periods in Fig. 6.3, and index of relative concentration (c), calculated as the ratio of the percentage of the subclass in the time period to the percentage of all globe patents in the time period

Subclass			1931–51		1957–72	
		Size	n	c	n	c
131	Terrestrial globe or accessory therefor	125	27	0.98	17	0.64
132	Relief globe	22	1	0.21	4	0.86
133	Having diverse use (e.g., pencil box, etc.)	22	5	1.03	2	0.43
134	Having magnet associated therewith	18	1	0.25	4	1.05
135	Having plural planar or curved surfaces (e.g., flat or frustoconical surfaces)	26	9	1.57	2	0.36
136	Rotated by mechanical drive	37	1	0.12	7	0.89
137	Collapsible or arranged for convenient assembly, disassembly, or storage	38	4	0.48	5	0.62
138	Inflatable	27	4	0.48	5	0.62
139	With means representing vehicle moving relative to earth	12	3	1.13	3	1.18
140	Space vehicle	37	0	0.00	36	4.48
141	With means indicating distance between points on earth	17	7	1.86	4	1.11
142	With means indicating time at different points on earth	50	11	1.00	7	0.66
143	With means demonstrating solar illumination of earth	39	9	1.04	5	0.60
144	With means demonstrating wind currents over earth	6	3	2.26	1	0.78
145	With internal light	53	24	2.05	5	0.44
146	With means to facilitate finding or reading indicia thereon	18	2	0.50	2	0.52
147	With map segment attachable thereto (e.g., continent, nation, etc.)	29	0	0.00	5	0.81
148	With suspension type support	15	1	0.30	1	0.31
Total number of patents within time period			105		101	

Because some patents were assigned to more than one subclass, the sum of the class sizes (591) exceeds the number of patents (475) in the dataset.

found no manufacturer who had licensed Lovegrove's patent, his strategy was aligned with the hydraulic pressing and stamping production methods adopted after World War I.

To better understand temporal trends in Fig. 6.3, I counted the patents in each subclass separately for the two periods of more frequent patenting.

These counts are listed in the columns labeled n in Table 6.2. Next, I calculated the percentage of each subclass's patents in the two eras. For example, of the 50 patents for globes "with means indicating time at different points on earth" (434/142), 11 (or 22 percent) were issued between 1931 and 1951, whereas of the 6 patents "with means demonstrating winds over earth" (434/144), 3 (or 50 percent) were issued during the same time period. To facilitate meaningful comparison, I divided each of these percentages by the 22.11, the percentage of the 475 patents in the dataset issued during the 1931–51 interval, which yields concentration indexes c of 1.00 and 2.26, respectively, indicating that time-difference globes were no more common during this period than in the dataset overall whereas wind-current globes had a relative concentration more than twice as high— the highest relative concentration of any subclass for 1931–51. Slightly less concentrated in this earlier era are internally lighted globes (434/138). By contrast, as the index 4.48 in the last column indicates, globes describing a space vehicle's movement "relative to earth" (434/140) were even more concentrated in the 1957–72 period, when they accounted for nearly 36 percent of the era's patents. Inflatable globes (434/138) and most other subclasses were not notably prominent during either era.

Because globes demonstrating wind currents are the least numerous subclass in the overall dataset, their strong relative concentration in the 1931–51 period—a concentration based on only three patents—seems meaningful only as a reflection of significant advances in atmospheric science spurred by the discovery of frontal boundaries and by improved means of exploring the upper atmosphere with aircraft, weather balloons, and radio transmitters.[47] As illustrated by the juxtaposition in Fig. 6.5 of drawings from the three patents, these innovations enhanced a more conventional terrestrial globe with elaborate three-dimensional devices describing air movement.

At least one of these patents seems grounded in utter fantasy. Seattle inventor Parvin Wright equipped his globe (Fig. 6.5, upper left) with an appliance apparently intended to relate the seasonal shift of the sun's elevation above the horizon to the shift in what climatologists call the Intertropical Convergence Zone (ITZ), where the trade winds converge.[48] Wright did not use conventional meteorological terminology, and how the device described in his drawings would portray this shift is not at all clear. No more obvious is the relevance of twin bands of cyclonic swirls to an understanding of weather and climate. Although Wright asserted, "According to a theory which I have evolved, and which is supported by considerable evidence, ethereal flows exist in the higher atmosphere,"

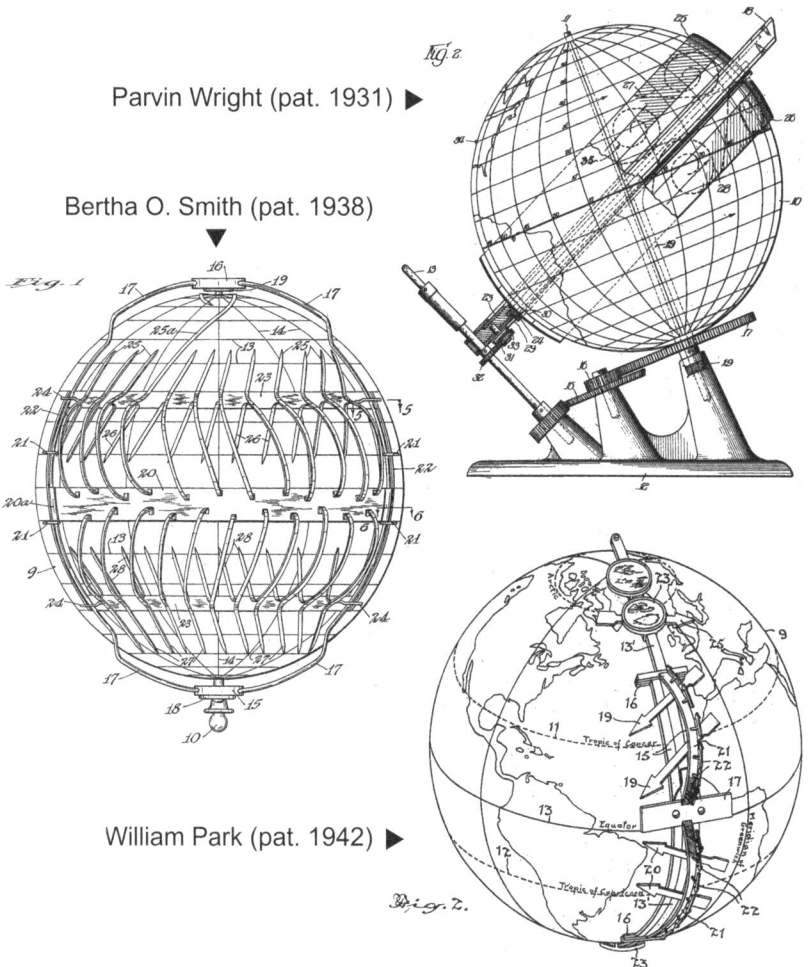

Parvin Wright (pat. 1931) ▶

Bertha O. Smith (pat. 1938) ▼

William Park (pat. 1942) ▶

Fig. 6.5 Key drawings from wind-flow globes patented by Parvin Wright, Bertha Smith, and William Park (*Upper right*: US Patent 1,836,423; 1931. *Left*: US Patent 2,105,619; 1938. *Lower right*: US Patent 2,305,894; 1942)

I found no evidence he ever published his insights in a meteorological journal. When he filed his application in 1927, the 71-year-old electrician already held several patents and no doubt understood both the patenting process and the value of persistence. Although the four years that lapsed before his application was approved might reflect some bafflement at the

Patent Office, examiners are not science editors adept at screening out crackpot ideas.

By contrast, Bertha Smith was better grounded in the basics of atmospheric circulation. As teachers of physical geography take pains to explain, the trade winds converge toward the ITZ from the northeast and the southeast to form a belt of comparatively stable air near the equator known as the doldrums, within which sailing ships were often becalmed for weeks. These converging surface winds reflect an upward movement of air, heated by intense solar radiation near the equator—air that then flows away from the ITC at greater altitude toward twin belts of descending air called the horse latitudes, one about 30° north of the equator and the other about 30° south. Climatologists call this pair of cyclic three-dimensional air currents Hadley cells, named for George Hadley, the British barrister who discovered them in the 1730s. These belts of descending air are also coupled with a pair of mid-latitude circulation cells, in which cooler air aloft descends at roughly 30°N and 30°S to feed westerly winds moving poleward near the surface.

Although tedious patentese obscures the science underlying Smith's globe, the principal atmospheric currents can be discerned in her primary drawing (Fig. 6.5, left) if one looks closely.[49] Multiple "flat metallic ribbons" attached to the doldrums band (labeled 20 on the drawing) extend outward from the surface of the globe and bend poleward, toward the horse latitudes (23), where each ribbon descends toward the surface before splitting in two, with one part pointing back toward the doldrums from the east (to portray the trade winds) and the other pointing poleward from the west (to portray the westerlies). Because Smith was well aware that the doldrums and horse latitude belts move seasonally, to follow the sun, her elaborate "wind indicating structure" is mounted on rods (17) running from pole to pole that allow a "limited upward and downward sliding movement."

When Smith applied for her patent in 1936, she was a 57-year-old Minneapolis resident who sold World Book encyclopedias—which might account for her interest in atmospheric circulation. Although Census records are inconclusive, thanks largely to a multitude of Smiths, she was born Bertha Olsen in Illinois, which probably accounts for the middle initial O, and had two years of college. Although she married at age 39, around 1918, I found no record of her husband. The 1930 Census found Smith living in an apartment with her aged parents, natives of Norway and Germany. In 1940 she was living by herself, in a hotel. It seems unlikely that she ever developed her patent because the globe would have been costly to produce and of too little pedagogic value to warrant the expense. She assigned half her patent rights to Anice Godfrey, a Minnesota portrait artist and photographer, who was a few years older. A faintly plausible explanation is that Anice might have helped with the drawings or volunteered her husband George, who ran a welding shop, to craft a model of the globe. In addition to the published patent, Smith's legacy is a single paragraph and a much reduced drawing in the *Official Gazette*.[50]

Toronto inventor William Park, who patented a less sophisticated embellishment for describing global winds in the early 1940s, left an even more obscure biographical footprint.[51] I know only that Park patented a planetarium a year later, and that he assigned half the patent rights for the globe, but not for his planetarium, to an equally elusive Toronto resident, Rudolf Dunbar.[52]

Not surprisingly, Park's planetarium partly resembles his globe, shown on one drawing with a large light bulb representing the sun and a smaller unlighted bulb for the moon. That several of the drawings for his globe patent include arrows depicting the northeasterly and southeasterly trade winds accounts for its assignment to the wind currents subclass (434/144). Elevated arrows (Fig. 6.5, lower right) that portray surface winds, not airflow aloft, can be shifted upward or downward along a single curved bar (13 in the drawing) to show seasonal influences. The bar runs from pole to pole and includes several "individual indicator discs … containing pictures representing the prevailing atmospheric conditions in the temperate zones"—useless trappings poorly placed. Unlike Smith's invention, which dramatized ascending and descending air currents, Park's contraption was less engaging than the juxtaposed January and July winds maps found in most physical geography textbooks.

Internally illuminated globes, only slightly less concentrated in the 1931–51 period than their climatological counterparts, underscore the light bulb's importance as both a source and a symbol of new ideas. With 24 patents issued during this period, subclass 434/145 reflects diverse ways of putting a light bulb inside a transparent or translucent globe. Variations include the bulb's size (large or small) or shape (round or elongated), its position within the sphere (in the center, at the top or bottom, or elsewhere near the edge), the number of bulbs (one or two), and diverse strategies for diffusing the light uniformly over the surface of the sphere.

None of these contrivances was as clever or instructive as the globe proposed in 1931, at the beginning of the period, by Herman Schulse, a

Wilmington, Delaware, mechanical engineer with several prior patents. In language clearer than that found in most patents, Schulse announced his invention as a "chronological and horological instrument" and described its goal concisely in a mere 85 words:[53]

> An object of the invention is to provide a clock driven replica of a terrestrial globe adapted to indicate automatically at all times, the position of the earth's shadow, the mean solar and sidereal time at any degree of longitude and to physically demonstrate the changing position of the earth relative to the sun with the changing seasons, whereby current information relative to the rising and setting of the sun and hours of light or darkness at any region of the globe, is always available.

His first drawing (Fig. 6.6, left) shows a translucent globe mounted on a frame that allows rotation about an axis inclined 23½° from the vertical. A clock at the base of the frame provides the time and is coupled to a

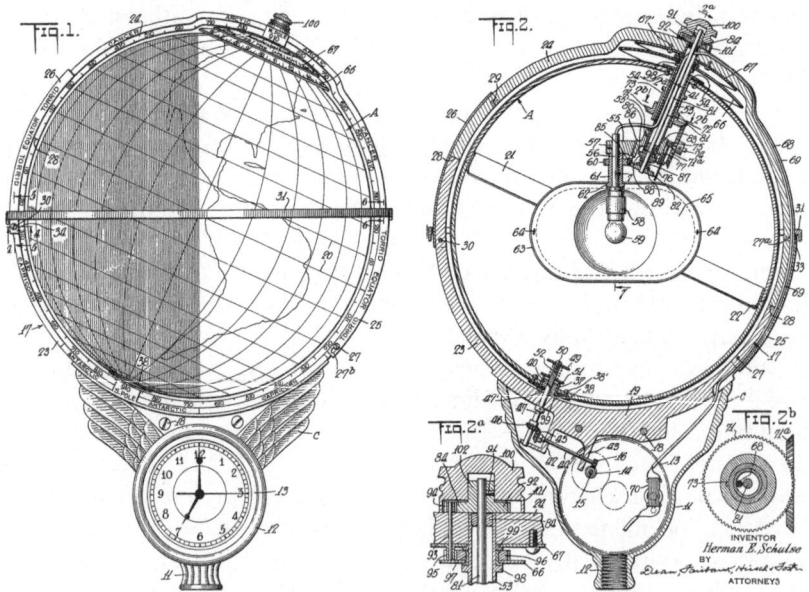

Fig. 6.6 The circle of illumination inside Herman Schulse's "Chronological Instrument" made a complete revolution in one year, while the translucent terrestrial globe rotated every 24 hours (US Patent 1,959,601; 1934)

transmission mechanism that moves the globe through a full 360° rotation once every 24 hours. At the center of the globe, a translucent filter (65 in Fig. 6.6, right) surrounding the bulb provides full illumination to half the globe and only as much light elsewhere as needed to read the meridians, parallels, coastlines, labels, and other features inscribed on the surface. To show the changing seasons, a set of gears near the top of the globe rotates the filter around the bulb a full 360° once every 365 days. The view in the left-hand drawing shows the pattern of illumination at the summer solstice, when locations north of the Arctic Circle have a full 24 hours of daylight. At the winter solstice, six months and a 180° turn of the filter later, only locations south of the Arctic Circle would be illuminated day and night. At the equinoxes, when the circle of illumination would intersect the poles, all places on the globe would rotate through exactly 12 hours of daylight.

Schulse knew how to run a business. When he registered for the draft in June 1917, at age 28, he reported his occupation as a "filter expert" at the Safety First Filter Co., of New York, where he was also a director, company treasurer, and inventor who had signed over several patents. By the late 1920s, he had patented three shoe-shining devices and assigned them to H. E. Schulse, Inc., which made the King Shiner Shoe Shine Station, one of which fetched $1159 at a January 2016 online auction.[54] Around the same time, he acquired an interest in globes, confirmed by a patent application filed in July 1929 for a non-lighted globe that merely rotated in synchronization with a clock—ironically this patent was issued in 1935, a year after the patent for his more complex illuminated globe, filed in 1931.[55] In the same year, he also filed seven design patents for ornate desk, floor, or wall-mounted globe frames. All of his globe patents were assigned to the Uniclox Corporation, located in Delaware, a friendly corporate home as well as Schulse's home state at the time.[56]

A 1930 news item in *School Science and Mathematics*, a periodical pitched to educators, confirms that Schulse did indeed develop his patent.[57] A press release probably accounts for the pronouncement that "Now man's ingenuity has created a mechanical world, a sort of miniature Mother Nature that goes through all its temporal antics before your eyes." The new instrument, readers were informed, was called the "Uniclok," to distinguish it from Uniclox, which owned the patent, and the Universal Clock and Globe Corporation, which handled marketing. The venture also involved Rand McNally, "whose sales agencies throughout the world provide international distribution for the new scientific instrument," and chemical giant (and Delaware neighbor) E. I. du Pont de Nemours, which manufactured the Viscoloid that formed the globe's translucent shell. Other sources attributed the clockwork and related gearing to Warren Telechron, an Ashland, Massachusetts, firm that had pioneered the electric clock.[58] Schulse understood the advantages of subcontracting to experienced outsiders. For an instruction manual, he compiled the 53-page *Uniclok Globe Handbook*, an introduction to astronomy that covers much more than how to reset the instrument after a power failure.[59]

Further confirmation that Schulse found at least a few buyers for his Uniclok is a photograph of the 30-pound floor model that an antiques dealer picked up around 2012 at a Tacoma, Washington, estate sale.[60] A cast iron frame 51 inches tall depicts a nude woman supporting a clock and a 12-inch globe. The light worked but the clock did not, which might account for a sale price of only $900. The original price of $265 is about $4000 in today's dollars.[61] A late 1931 ad in *The Literary Digest* offered an "automatic desk model" for $160 and "hand operated" desk and floor models for $90 and $115.[62] The rarity of Schulse's globes at least partly reflects the onset of the Great Depression, which further diminished the small upscale market for expensive toys.

Artificial satellites provided another opportunity for inventors to patent globes that mixed science and play. The Soviets launched Sputnik into orbit on 4 October 1957, and by the end of the decade, inventors had filed 17 ultimately successful patents for globes in the "Space vehicle" subclass (434/140). By 1972 the number of patents in this category had grown to 36, passing all other categories in the 1957–72 cluster (Table 6.2).[63] Fifteen of these patents demonstrated the importance of amusement by unabashedly incorporating *toy* in their specification, 9 of these 15 even self-identified as toys in their title, and another three reported *amusement* or *recreation* as an important or sole purpose.

I doubt that any of these inventions ever appeared in toy stores. Though clever, they lacked play value: the capacity to hold the interest of a child or adult for more than a few minutes. Take, for example, the "Toy Earth Satellite" patented by Lambert Decker of Ulster Park, New York.[64] The satellite is a small metal ball (35 in Fig. 6.7) held in place by a suitably powerful magnet (34) inside "a hollow globe 15 of nonmetallic material." The ball/satellite orbits the globe as the child (or adult?) rotates the crank (20). That's it. The most fun would be to crank the apparatus so rapidly that the ball flies off the globe—possibly hitting someone in the eye. The likelihood of lawsuits is one of several explanations for why Decker's

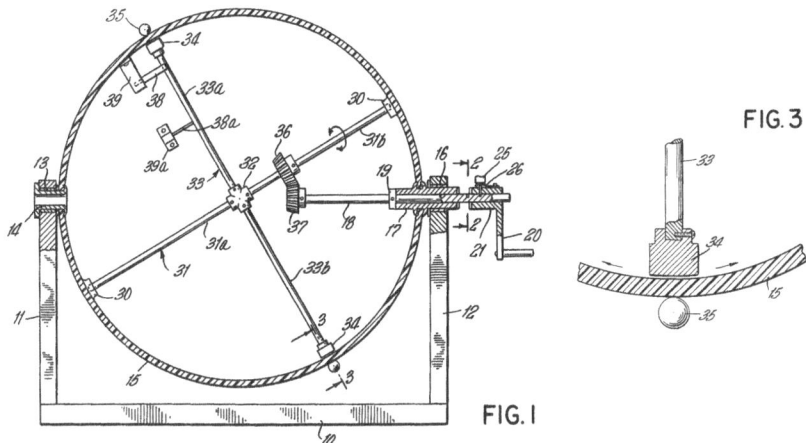

Fig. 6.7 Lambert Decker's toy globe and satellite (US Patent 2,870,550; 1959)

invention never made it to market. Consumer safety was not a responsibility of the Patent Office.

To make a space vehicle toy more interesting, the inventor had to get the rocket off the ground and keep it on a controlled flight path. To meet this challenge, Kalman Benko, of Cleveland, Ohio, placed a rocket cum satellite (1 in Fig. 6.8) inside plastic tubing held in place by struts.[65] Air from a blower in the base of the globe forced the vehicle to circle the earth before going into a single orbit around "moon 28"—a flagrant exaggeration of scale insofar as earth's diameter is four times that of the moon and the intervening distance is roughly 30 times earth's diameter. An air control valve described in another drawing allowed air to escape in advance of the rocket and also let the space vehicle either drop back toward earth or make another figure-eight. Despite this interactivity, Benko's cumbersome toy never approached the play value of a simple electric train set. But at least it wouldn't put someone's eye out.

Although Decker's and Benko's toys could no doubt be made to work, the fanciful invention patented by brothers Thomas and Frank Novak, of Brownsville, Pennsylvania, might have made cartoonist Goldberg scratch his head.[66] Like Benko, the Novaks relied on forced air from a pump below the globe. A cross section of their apparatus (Fig. 6.9, left) shows how an "upward air blast" from one end of rotary air duct 14 would keep "sphere 21 … made from a suitably light material such as styrofoam" a constant

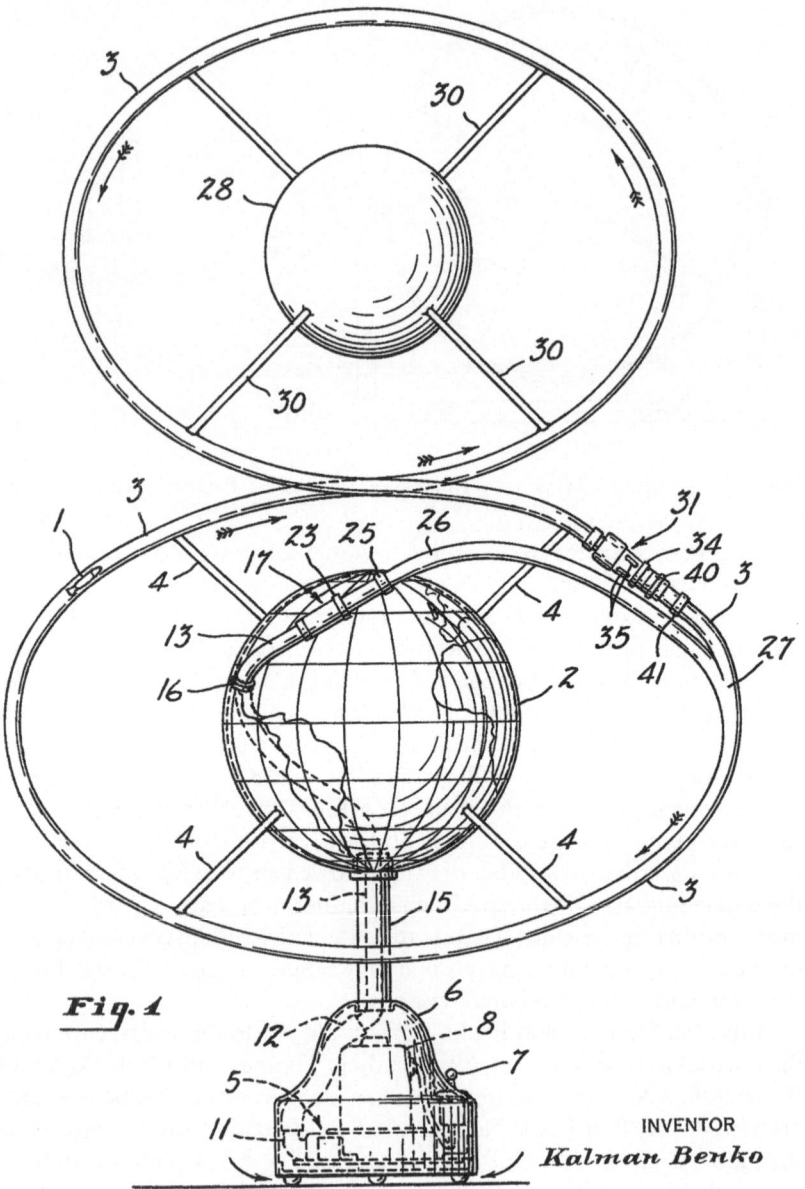

Fig. 6.8 Kalman Benko's "Earth Circling Satellite Toy" included a circuit around the moon (US Patent 2,890,537; 1959)

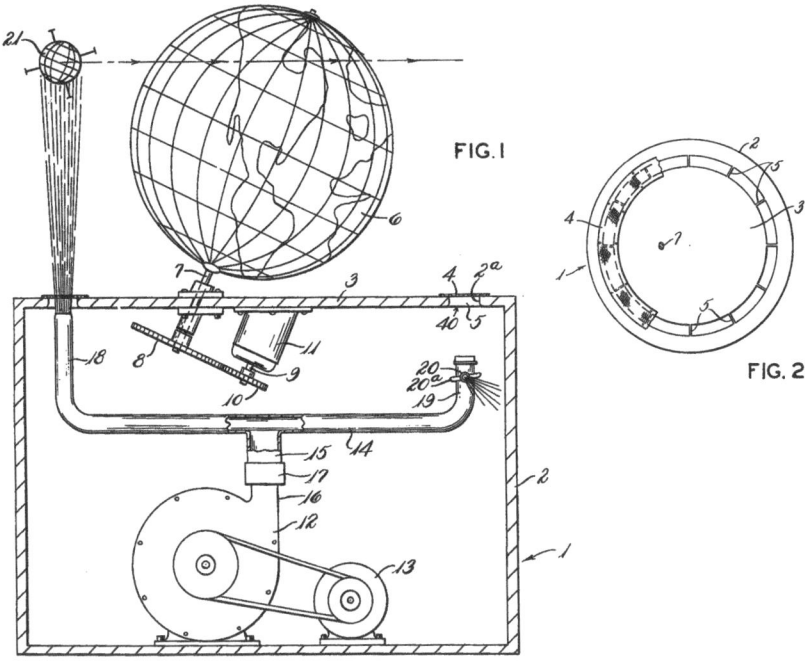

Fig. 6.9 Thomas and Frank Novak linked a stationary globe to a moving satellite held aloft by a current of air (US Patent 3,083,497; 1963)

distance above platform 3 while air from "jet nozzle 20," at the other end, turned the air duct in a horizontal plane. A smaller plan view (Fig. 6.9, right) describes circular screen 4, designed to let the vertical air blast pass upward through the platform. Would it not have been simpler, and more reliable, to anchor the satellite to a metal rod made to revolve around the globe by an electric motor?

Turbulence and gravity undermined Novaks's invention. Unless the satellite was tethered, adjusting the vertical air current to keep it at the right height would have been nearly impossible, particularly so because of the air current's rotary motion. Why their device was more mechanical than electronic is a mystery insofar as both brothers had studied electronics and worked in a family business that sold and serviced televisions.[67]

Thomas was older, which might explain his name appearing first on the patent, but Frank was probably the clever one. In 1965, four years after applying for the patent, he entered the Gravity Research Foundation's annual essay contest, which drew 138 submissions. Although Frank was not a finalist—his entry, "Gravity—Self Excitable and Unipolar," rambled on about "gravity energy particles" and the "unlimited power" of "gravity shields"—his short biography described an "inventor [and] amateur scientist [who had] experimented in magnetic actions to high frequency voltages and electro static voltages, ions and ionic motors [and] also air currents which cause a satellite toy to orbit a fixed globe."[68] The essay, like the patent, was a way to assert achievement.

A few 343/140 patents were much too intricate to be considered toys, even for the scientifically inclined rich. For example, the "Method and Means for Displaying Positions and Motions of Objects, in Relation to the Earth," patented in 1961 by Edward J. Madden, of Alexandria, Virginia, looks like a display system for Defense Department, CIA, or White House briefings (Fig. 6.10)—and perhaps was, insofar as all three agencies are close to Alexandria, where Madden flew below the radar of city directories.[69] It might well be an example of clandestine Cold War–era geospatial innovations that, as historian of technology John Cloud noted, were "hidden by being carefully concealed in plain view."[70] The artwork's highly generalized treatment of the Caribbean—note the exaggerated and capriciously displaced Cuba (middle right in Fig. 6.7)—suggests a pressing need for satellite surveillance in the lead-up to the Cuban Missile Crisis of 1962.

Two subclasses not prominent during the 1931–51 and 1957–72 high-patenting periods are also noteworthy. "Terrestrial globe ... having diverse use (e.g. pencil box, etc.)" (434/133) was a catchall for inventions that found a use for the otherwise empty space inside a globe. Examples include a "toy money safe," a "combined globe and bank," a fish aquarium, a lampshade, an ashtray, and a radio cabinet.[71] More geographically relevant is the "Educational Globe" (Fig. 6.11) patented in 1889 by New York City lithographer Olin D. Gray, who proposed a small sphere containing a Z-fold strip of flexible material bearing "on one of its faces pictures indicative of historical epochs since the discovery of the Western Hemisphere, and on its reverse face representations of buildings of the 'World's Fair'."[72] I found no evidence that this promising tourist souvenir was ever manufactured; it was Gray's only patent.

By contrast, "Terrestrial globe ... with means to facilitate finding or reading indicia thereon" (434/146) includes an ingenious device that packed a wealth of location data inside an 18-inch floor globe. Its inventor, Charles M. Williams, was a Los Angeles resident who might have

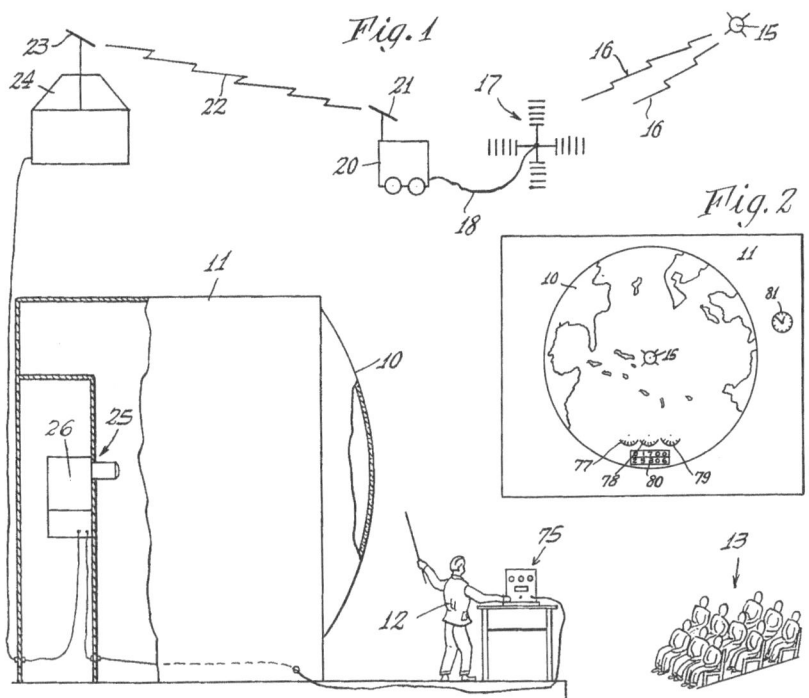

Fig. 6.10 Edward Madden's system for reporting a satellite's position in real time was not representative of most inventions in its terrestrial globe subclass for space satellites. Icon 15, at the center of the display (*right middle*), pinpoints the satellite's position on the globe (US Patent 3,003,257; 1961)

held earlier patents for a mop head and a mousetrap—at least two people with the same name lived in L.A. in the early 1900s, and city directories indicate they moved frequently.[73] In 1922, he filed a patent application for a "Geographical Globe" with an internal index to thousands of places. Inside the globe (Fig. 6.12), a pair of spools 34 and 38 anchor a long, thin tape inscribed with the places' names, latitudes, and longitudes.[74] The names are ordered alphabetically, and the user advanced the tape beneath apertures 22 and 23 by turning hand wheel 31. Each aperture had a small magnifying lens that made the tiny printing legible.

Unlike Gray, Williams manufactured and actively promoted his invention, which thrived in the mid-1920s. He obtained endorsements from

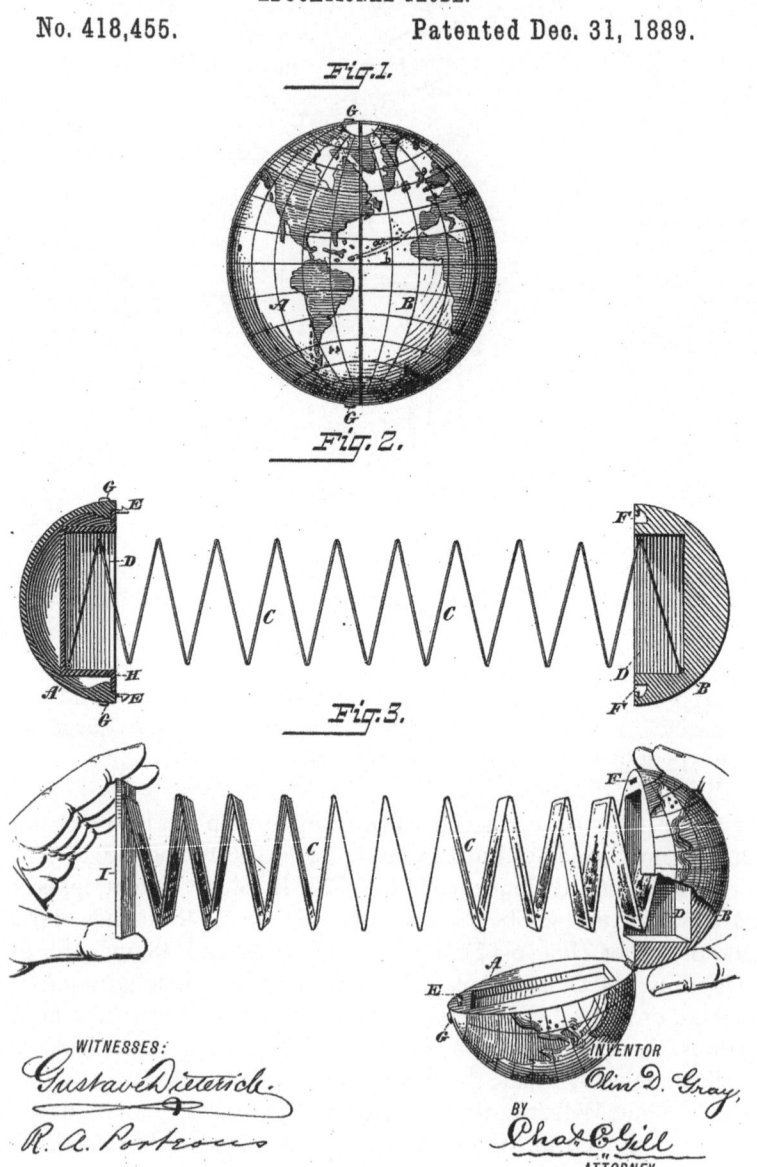

Fig. 6.11 Olin Gray's patent described a small globe containing pictures on a simple accordion fold (US Patent 418,455; 1889)

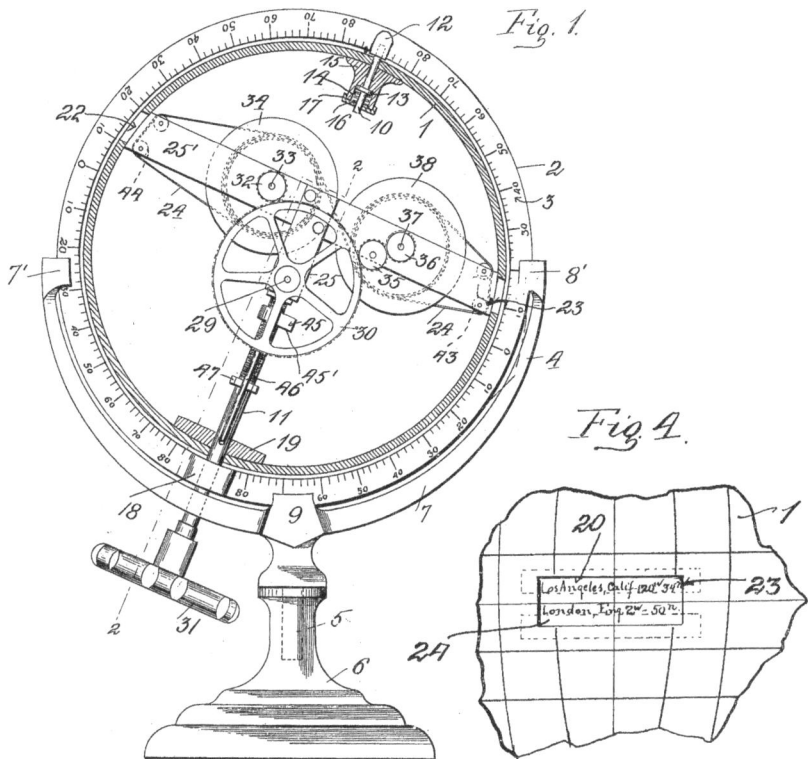

Fig. 6.12 Complex gearing advanced a long, thin tape with place names and coordinates beneath viewing apertures on opposite sides of Charles Williams's Index Globe (US Patent 1,511,487; 1924)

Alexander Graham Bell, several government agencies, the National Education Association, and the National Geographic Society, with which he hoped to partner. In late 1919, while visiting Washington, DC, Williams described his frustration in a letter to Clark University geography professor Wallace Atwood. He had proposed manufacturing 12-inch globes with 15,000 names for $3.50 and selling them for $12.00, leaving ample profit to recover the initial investment. The obstacle was Gilbert Grosvenor, editor of *National Geographic Magazine* and imminent president of the Society. "Everybody here has expressed themselves as delighted, except Grosvenor can't come down to sign up, [and] acts more like a brainless

girl trying to buy a spring hat."[75] Williams eventually found another partner and set up a factory in California.

In 1920 the *Los Angeles Times* had applauded him as a local genius
who might make geography "popular even among school children."[76] His
"index globe" was particularly timely because "the World War [had] made
junk of all the millions of globes in use in schools, colleges, and homes."
The globe had three rectangular windows situated over the comparatively
empty oceans. One was a portal to 30,000 place names; another window
addressed "several thousand pictures on a reel," and a third offered "history
and important facts" about various cities and countries. A later refinement
was a single 51-foot index strip, covering 43,000 places with 200,000 words
printed 28 lines to the inch on a specially designed printing press.[77] The patent's claims were sufficiently broad to cover these modifications.

Like Herman Schulse's illuminated chronometer/globe, Williams's
18-inch floor globe was a decidedly upmarket device, priced at $210–225,
depending on the style and height of the frame.[78] And like Schulse's
enterprise, Williams's firm was devastated by the Depression, and all of
its machinery, unsold globes, and miscellaneous parts were sold at auction
in October 1933. As advertised in the *Los Angeles Times*, the inventory
included "8000 Map Covers for 12″ Globes," suggesting that the market
for the smaller, less expensive model had also collapsed.[79]

Because most patented globes were not commercially sustainable, they
represent a much wider range of development of the globe as a decorative
object, an educational toy, and a device for demonstrating differences in
time and solar illumination than do the globes typically found in museums
and map collections or examined in the literature on map history. Though
many (if not most) patented globes are too complicated, too narrow in
function, or simply too silly for commercial production, collectively these
largely hypothetical but nonetheless fascinating contraptions warrant a
more detailed examination than possible here. In their exceptionally broad
diversity, patented globes might be the epitome of unknown, unacknowl-
edged cartographic innovation.

NOTES

1. H. Ariel Norris, "Art of Hydrographic Surveying," US Patent
   1,662, issued 27 June 1840.
2. Robert Piggot, "Apparatus for Teaching Geography and
   Astrography," US Patent 2,426, issued 17 January 1842. The book

mentioned in Morris's headnote might have been disassembled in the early twenty-first century when published patents were scanned for the Google Patents project. Robert Beebe, archivist, National Archives at Kansas City, oral communication, 11 January 2016.

3. Until the late nineteenth century, handwritten patent applications were folded into quarters and put into legal-size envelopes, for filing. Beebe, oral communication. Moreover, because the patent system was not fully developed, the archived files for patents issued before around 1870 are often incomplete. See Nathan Reingold, "U. S. Patent Office Records as Sources for the History of Invention and Technological Property," *Technology and Culture* 1 (1960): 156–67.

4. A contemporary listing of engravers praised Piggot's talent, listed significant engraved portraits of James Monroe and ten other prominent persons, and lamented his change of occupation. William Spohn Baker, *American Engravers and Their Works* (Philadelphia: Gebbie & Barrie, 1875), 135–37.

5. Pamela A. Hartman, *A History of the Western High School in Seven Decades, 1844-1913* (Baltimore: Fleet-McGinley, 1915), 11.

6. Mantle Fielding and David McNeely Stauffer, *American Engravers Upon Copper and Steel* (Philadelphia: Burt Franklin, 1917), 212.

7. The absence of an entry for Piggot in a comprehensive survey of early American globes suggests his globe was not manufactured in significant numbers, if at all. See Ena L. Yonge, *A Catalogue of Early Globes, Made Prior to 1850 and Conserved in the United States* (New York: American Geographical Society, 1968).

8. Silas Cornell, "Mounting Globes," US Patent 4,098, issued 5 July 1845.

9. For examples, see the Rochester Photo Images database, available online from the Monroe County Library System, at http://photo.libraryweb.org/carlweb/jsp/photosearch_pageADV.jsp.

10. Silas Cornell, *A Description of Silas Cornell's Improved Terrestrial Globe with the Manner of Using It: Intended for the Use of Schools, Academies, and Families,* 2nd ed. (Rochester, 1845). The title page notes that it was "published by the author; printed by Canfield & Warren," which also published Cornell's city street maps.

11. Cornell, *Description*, vii.

12. "S. Cornell's District School Globe," *District School Journal of the State of New York* 5.4 (July 1944): 126.

13. Galileo [pseud.], "New Apparatus for Teaching Geography," *Common School Journal* 8.21 (November 1846): 334–35.

14. "Terrestrial Globe, Cornell, Silas," David Rumsey Map Collection, http://www.davidrumsey.com/maps6354.html.

15. Kenneth E. Behring Center, National Museum of American History, Smithsonian Institution, "Cornell 9-Inch Terrestrial Globe," Digital Public Library of America, http://collections.si.edu/search/results. htm?q=record_ID%3Anmah_1064464&repo=DPLA.

16. Yonge, *A Catalog of Early Globes*, 22.

17. Nordis Felland, "Ena L. Yonge, 1895–1971," *Geographical Review* 62 (1972): 414–17.

18. Ellen E. Fitz, *Hand-Book of the Terrestrial Globe, or, Guide to Fitz's New Method of Mounting and Operating Globes, Designed for the Use of Families, Schools, and Academies* (Boston: Ginn Brothers, 1876).

19. A January 2016 search on Amazon.com found over a half-dozen offerings released between 2008 and 2015 by BiblioBazaar, Cornell University Library, Forgotten Books, Leopold Classic Library, and RareBooksClub.com, among others. In addition to hardcover and softcover editions, HardPress Publishing was selling an electronic version, compatible with Amazon's Kindle reader.

20. Ellen E. Fitz, "Improvement in Globes," US Patent 158,581, filed 9 December 1874, and issued 12 January 1875.

21. Judith Tyner, "The Hidden Cartographers: Women in Mapmaking," *Mercator's World* 2.6 (November/December 1997): 46–51.

22. The sketchy evidence of Asa's employment includes the 1850 Census, which lists his occupation as editor, and an advertisement in an 1864 issue (volume 17, no. 6) of *The Massachusetts Teacher: A Journal of School and Home Education*, which lists him as author of *American School Hymn Book* and *School Exhibition Book*.

23. William B. Fowle and Asa Fitz, *An Elementary Geography for Massachusetts Children* (Boston: Fowle and Capen, 1845), 5–7.

24. Fitz, *Hand-Book*, 57.

25. Ellen E. Fitz, "Mounting and Attachment for Terrestrial Globes," US Patent RE 9,557, originally issued 12 January 1875, reissue application filed 29 May 1879, and issued 8 February 1881.

26. Reissue patents are discussed briefly, in context, in Reingold, "U. S. Patent Office Records," 161, 163.

27. Ellen E. Fitz, "Globe," US Patent 263,886, filed 5 November 1881, and issued 5 September 1882. Also see Deborah Jean

Warner, "The Geography of Heaven and Earth," *Rittenhouse: Journal of the American Scientific Instrument Enterprise* 2.6 (1987): 52–62; esp. 62.

28. Massachusetts vital records report that Fitz died in Watertown, MA, on 12 October 1886. Ancestry.com, *Massachusetts Town and Vital Records,* 1620–1988.

29. Bonham's, New York, "Tabletop Globe: Fitz, Ellen Eliza," 22 October 2014, https://www.bonhams.com/auctions/22247/lot/26/.

30. See, for example, Autumn Stanley, *Mothers and Daughters of Invention: Notes for a Revised History of Technology* (Metuchen, NJ: Scarecrow Press, 1993), 452; Judith Tyner, "The Hidden Cartographers"; and Will C Van den Hoonaard, *Map Worlds: a History of Women in Cartography* (Waterloo, Ontario, Canada: Wilfrid Laurier University Press, 2013), 63.

31. Deborah Jean Warner, "The Geography of Heaven and Earth," *Rittenhouse: Journal of the American Scientific Instrument Enterprise* 2.8 (1987): 108–37; quotation on 118.

32. US Patent and Trademark Office, "December 15th Marks the 165th Anniversary of The Great Patent Office Fire of 1836; Flames Destroyed a Significant Period of U.S. Patent History" [press release], 14 December 2001, http://www.uspto.gov/web/offices/com/speeches/01-60.htm.

33. "X patent," USPTO Glossary [online], http://www.uspto.gov/main/glossary/#x.

34. Elizabeth Oram, "Globe for Teaching Geography," US Patent 6,337X, 12 January 1831, Directory of American Tool and Machinery Patents, http://www.datamp.org/patents/display-Patent.php?number=6337&typeCode=3.

35. *A Digest of Patents, Issued by the United States, from 1790 to January 1, 1839: Published by Act of Congress, under the Superintendence of the Commissioner of Patents, Henry L. Ellsworth* (Washington, DC: Peter Force, 1840), 410, 412.

36. "23. For an Instrument for the Teaching of Geography; Elizabeth Oram, City of New York, January 12," *Journal of the Franklin Institute* n.s., 7 (May 1831): 307–8.

37. Roger Sherman Skinner's *New York State Register for the Year of Our Lord 1830* ([New York: Clayton & Van Norden, 1830], 298), lists three principals for New York High School: two for the Male High School and one (Elizabeth Oram) for the Female High

School. *Documents of the Board of Education of the City of New York, for the Year ending December 31, 1860* ([New York: George Russell, 1861], 6) includes *Oram's Grammar*, priced at 25 cents "per single copy," in a list of 13 grammar books.

38. Henry B. Goodyear, "Method of Relieving Geographical Outlines on Molded Elastic Globes," US Patent 31,311, issued 5 February 1861.

39. Jan Mokre, "Globe: Cultural and Social Significance of Globes," in *Cartography in the Twentieth Century* (Vol. 6 of *The History of Cartography*), ed. Mark Monmonier, 558–63 (Chicago: University of Chicago Press, 2015).

40. Gorham D. Abbot, "Improvement in Globes," US Patent 80,891, issued 11 August 1868. Also see "Meeting at Mrs. Hopkins's Home to Advance the Project—The Interesting Career of Gorham D. Abbot," *New York Tribune*, 19 May 1899, 5; and "Opinion in the Case of the American Hard Rubber Company," *New York Times*, 4 June 1861.

41. John Tallis, *Tallis's History and Description of the Crystal Palace, and the Exhibition of the World's Industry in 1851* (London, 1852), 2:246.

42. Goodyear, "Method of Relieving Geographical Outlines on Molded Plastic Globes."

43. Edward H. Knight, ed., "Globe," in his *Knight's American Mechanical Dictionary*, 2:986–88, esp. 987 (New York: J. B Ford and Company, 1875).

44. Jan Mokre, "Globe: Manufacture of Globes," in *Cartography in the Twentieth Century* (Vol. 6 of *The History of Cartography*), ed. Mark Monmonier, 563–65 (Chicago: University of Chicago Press, 2015). Also see Dianne Lee van der Reyden, "History, Technology and Care of Globes: Case Study on the Technology and Conservation Treatment of Two Nineteenth-Century Time Globes," Smithsonian Center for Materials Research and Education, 1986, https://www.si.edu/mci/downloads/RELACT/globes.pdf.

45. Francis Augustus Lovegrove, "Printing Machine for Spheres and the Like," US Patent 1,728,351, filed 10 August 1928, and issued 17 September 1929.

46. Francis Augustus Lovegrove, "Pipe Threading and Cutting Machine," US Patent 1,224,996, filed 21 March 1917, and issued

30 October 1917; and Francis A. Lovegrove, "Signal for Submarine Vessels," US Patent 1,113,799, filed 10 October 1913, and issued 13 October 1914.

47. Mark Monmonier, *Air Apparent: How Meteorologists Learned to Map, Predict, and Dramatize Weather* (Chicago: University of Chicago Press, 1999), 57–80.

48. Parvin Wright, "Instrument for Locating Storm Zones," US Patent 1,836,423, filed 31 December 1927, and issued 15 December 1931.

49. Bertha O. Smith, "Device for Depicting Prevailing Air Currents of the Earth," US Patent 2,105,619, filed 22 June 1936, and issued 18 January 1938.

50. Bertha O. Smith, "2,105,619—Device for Depicting Prevailing Air Currents of the Earth," *Official Gazette of the United States Patent Office* 486 (18 January 1938): 530–31.

51. William Park, "Tellurian," US Patent 2,305,894, filed 28 October 1940, and issued 22 December 1942.

52. William Park, "Planetarium," US Patent 2,318,961, filed 20 September 1941, and issued 11 May 1943.

53. Herman E. Schulse, "Chronological Instrument," US Patent 1,959,601, filed 19 September 1931, and issued 22 May 1934.

54. Schulze's three patents were "Shoe Shining Installation," US Patent 1,688,753, filed 29 October 1927, and issued 23 October 1928; "Shoe Shining Installation and Elements Thereof," US Patent 1,828,820, filed 17 January 1927, and issued 27 October 1931; and "Shoe Shine Last," filed 5 June 1928, and issued 24 May 1932. For the auction result, see Morphy Auctions, "Lot #: 785 - King Shiner Shoe Shine Station," sold 30 January 2016, http://morphyauctions.hibid.com/lot/31120-69695-202472/king-shiner-shoe-shine-station/.

55. Herman E. Schulse, "Chronological Instrument," US Patent 2,000,457, filed 16 July 1929, and issued 7 May 1935.

56. Schulse's design patents were numbered 84,794, 84,795, 84,796, 84,903, 85,206, 85,207, and 86,937.

57. "The Uniclok," *School Science and Mathematics* 30 (1930): 443.

58. See, for example, the searching aid for the Joseph Downs Collection of Manuscripts and Printed Ephemera at the Winterthur Library, which describes drawings of the Uniclok trademark and views of the globe from multiple angles and also notes that "Warren

Telechron Co. made [the instrument] for the Universal Clock and Globe Corp." Series II: Drawings and blueprints, Box 2, Folder 2, online at http://www.winterthur.org/html/downs_collection_ and_winterthur_archives/xhtml/JDCMcKinstryFM.htm.

59. Herman E. Schulse, compiler, *The Earth: Uniclok Handbook: A Pocket Manual of the Earth, Moon, Sun, Planets, Stars* (Wilmington, DE: Universal Clock and Globe Corp., 1931). In his copyright registration, Schulse listed himself as the work's compiler. *Catalogue of Copyright Entries*, n.s., 28,6 (1931): 824.

60. "Sensational 1940s [sic] Figural Floor Globe Clock made by Telechron," 1stdibs online marketplace, https://www.1stdibs. com/furniture/more-furniture-collectibles/globes/sensational-1940s-figural-floor-globe-clock-made-telechron/id-f_739514/ [accessed 18 February 2016].

61. According to the US Inflation Calculator, a Consumer Price Index application online at http://www.usinflationcalculator.com/, a product that sold for $265 in 1931 would cost $4,061.42 in 2013.

62. Uniclok [advertisement], *Literary Digest* 111 (14 November 1931): 47.

63. Two patents, one filed in late 1956 (3,003,257) and the other in early 1957 (2,825,151), predated the Soviet launch—hardly surprising because the Space Race was well underway.

64. Lambert Decker, "Toy Earth Satellite," US Patent 2,870,550, filed 28 January 1958, and issued 27 January 1959.

65. Kalman Benko, "Earth Circling Satellite Toy," US Patent 2,890,537, filed 8 September 1958 and issued 16 June 1959.

66. Thomas Novak and Frank Novak, "Satellite Toy, Display Article, or the Like," US Patent 3,083,497, filed 19 March 1961, and issued 2 April 1963.

67. "Thomas Novak" [obituary], *Uniontown (PA) Herald-Standard*, 22 February 2005; and Gravity Research Foundation, Entries Received, 1965, http://gravityresearchfoundation.org/pdf/ awarded/1965/1965authors.pdf.

68. Frank Novak, "Gravity—Self Excitable and Unipolar," www.gravi tyresearchfoundation.org/pdf/awarded/1965/novak.pdf.

69. Edward J. Madden, "Method and Means for Displaying Positions and Motions of Objects, in Relation to the Earth," US Patent 3,003,257, filed 1 November 1956, and issued 10 October 1961.

70. John Cloud, "American Cartographic Transformations during the Cold War," *Cartography and Geographic Information Science* 29 (2002): 261–82; quotation on 279.

71. See United States patents D19,140; 1,196,108; 4,272,372; 649,079; 2,341,092; and 2,138,959.

72. Olin D. Gray, "Educational Globe," US Patent 418,455, filed 18 November 1889, and issued 31 December 1889.

73. Charles M. Williams, "Mop Head," US Patent 635,216, filed 4 January 1899, and issued 17 October 1899; and Charles M. Williams, "Animal Trap," US Patent 640,890, filed 4 January 1899, and issued 9 January 1900.

74. Charles M. Williams, "Geographical Globe," US Patent 1,511,487, filed 29 March 1922, and issued 14 October 1924.

75. Charles M. Williams to Wallace W. Atwood, 26 December 1919. Clark University Archives, The Dr. Wallace Atwood Papers, series 1, box B4-1-4.

76. "Makes Geography Easy: Local Man's Globe Tells Everything," *Los Angeles Times*, 6 June 1920, 117.

77. "Invents Combination Globe and Atlas," *Popular Science Monthly* 116.4 (April 1930): 34; and Ransome Sutton, "What's New in Science: A Marvelous Memory Machine," *Los Angeles Times*, 20 September 1931, K17.

78. Prices from images of a 1920s-era catalog of the Williams-Pridham Index Globe, advertised on eBay and sold 17 May 2013. http://www.worthpoint.com/worthopedia/williams-pridham-index-globe-440937091 (accessed 1 March 2016).

79. "Auction … Bankrupt Stock—Machinery and Equipment of Williams-Pridham Index Globe Co.," *Los Angeles Times*, 8 October 1933, part I, p. 10.

CHAPTER 7

Current Events

Before digital computing radically transformed mapping and map use, inventors had devised clever ways for using electricity to illuminate the whole map or individual symbols as well as make maps interactive. This final chapter explores the development of electrically operated cartographic inventions before the emergence of software-driven electronic cartography in the late 1950s.[1] It begins by looking at pre-software cartographic applications of electricity and concludes with the story of how a so-called patent troll used a simple mall map to threaten giants of the geospatial industry with a frivolous lawsuit. As with occasional slipups in scholarly publishing, patent examiners sometimes fail to apply strict standards of novelty and non-obviousness.

My search for noteworthy early examples of electronic cartography began with the 283/34–35 subclasses examined in previous chapters. While assigning patents to a specific category such as travel or folding, I tagged 15 patents in which an electric light or circuit seemed like a key part of an invention for which the cartographic image came from a printing press. Excluding eight patents filed after 1959 and a 1941 invention that merely backlit a map on glass for reading in the dark left a core group of six Printed matter maps.[2]

Wary that I-know-it-when-I-see-it tagging might have missed some noteworthy innovations, I also searched the database for additional patents with *electric* anywhere in the text, and found two mechanical route

© Mark Monmonier 2017
Mark Monmonier, *Patents and Cartographic Inventions*,
Palgrave Studies in the History of Science and Technology,
DOI 10.1007/978-3-319-51040-8_7

guides discussed in Chap. 3. Neither was quintessentially electrical: Pio Papini noted that his drawings described a "purely mechanical means," preferred because an "electrical mechanism is very apt to get out of order especially if worked in the open air where it may be affected by dust, dirt or lubricating oil," and Henry Hubschmitt described a backlit nonconductive ribbon map with openings appropriately located so that "contact fingers" on opposite sides could close the circuit between a battery and a buzzer, thereby alerting the motorist to an approaching turn.[3] My text search also turned up an additional 11 patents, appropriately ignored because they described a simplistic front- or backlit illuminated map, mentioned electric railways or utility lines as part of a map's content, or used static electricity to hold transparent maps in place.[4] The backlit maps were no more quintessentially electrical than the illuminated globes discussed in the preceding chapter.

Earliest among the core group of six was the "Directory-Board" patented by Swiss inventor August Merk-Wirz, whose first drawing (Fig. 7.1, upper) juxtaposes a list of tourist attractions, theaters, department stores, and restaurants with a map of a hypothetical city center.[5] Letter codes on the map (e.g., HP for the Hotel Post, N for national monument) identify key locations, listed in a menu on the left. Next to each item is a pushbutton linked electrically to a light on the map. Pushing the button for the R. Wertheim department store, for example, would light a bulb behind the corresponding translucent symbol for the building labeled RW.

A circuit diagram with four buttons (11 in Fig. 7.1, middle) and their corresponding lights (10) describes the principle: pushing one of the buttons not only completes the circuit through the corresponding light bulb on the map but also sends current through a differently colored bulb (12) representing the directory's location—a mall map's "You are here" symbol. A battery (17) provides current for both bulbs, temporarily wired in series, like a string of single-wire Christmas-tree lights. The bottom drawing is a cross section with the pushbutton on the left and the selected lamp (10 in Fig. 7.1, lower) on the right, just beneath a narrow break (18) in the sheet of opaque paper (6) on which the map is printed. Which explains the patent's assignment to the Printed matter/Maps category.

Merk-Wirz aptly described his invention as an improvement to directory maps like those placed in kiosks to guide people new to the area: a common cartographic genre in early-twentieth-century European cities. Because a typical directory map included many more destinations than its schematic equivalent in the patent drawing, finding a particular

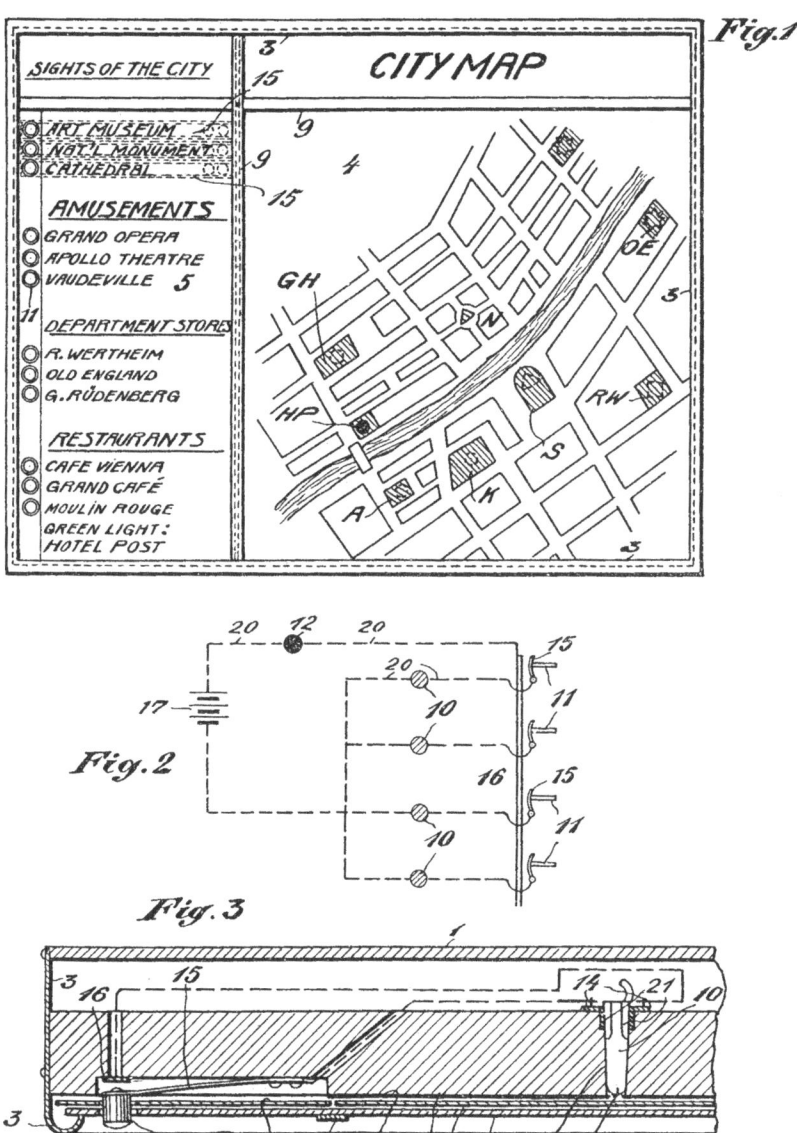

Fig. 7.1 August Merk-Wirz invented a forerunner of the pushbutton mall map (US Patent 1,132,108; 1915)

destination amid a forest of densely packed names and symbols was not always straightforward, even with a grid of cells referenced by letter and number. Merk-Wirz's invention not only pinpointed the destination's location but also addressed the question of "how far?" by simultaneously highlighting the directory's location. The patent's specification suggested a wider use of the electric-directory principle for maps of amusement parks, exposition grounds, and railway lines.

A resident of Zurich, Merk-Wirz had patented his invention in other countries, including Canada, France, and Great Britain. For his American filing, he chose a Chicago patent attorney, Berthold Singer, who helped him craft 19 claims, which the Patent Office ultimately whittled down to 9.

Filed in mid-June 1913, his application met stiff resistance from an examiner, Sidney Smith, who sensed similarity to a patent granted two years earlier to New Orleans resident Edwin Powell for an electric "room indicator" intended to help hotel clerks determine quickly whether a particular room included a bath or was already taken.[6] Singer convinced Smith that Powell's floor plan—a map of sort but included only as an "alternative form"—was operationally different from Merk-Wirz's city map, "adapted for use by any party" rather than just one user, the hotel clerk.[7] Smith had also rejected the application's first claim because of similarity to an 1892 patent for a non-electric pin map intended to help a salesman with clients scattered across a broad area: a rejection that Singer dealt with quickly by cancelling the claim.[8] At times the vetting of a patent resembles a game in which the inventor asserts overly broad claims, and the examiner counters with questionable grounds for rejection. Win some, lose some.

Even the most basic facts of Merk-Wirz's education and occupation are elusive, as is evidence that his directory board was ever manufactured in the United States. In Great Britain, by contrast, court records indicate that his invention was licensed to The Silent Guides, Ltd., which had installed electric map directories in several Underground stations, in possible conflict with a similar device that another inventor, Malcolm Quelch, now sought to patent. In 1923 Quelch petitioned to have Merk-Wirz's British patent revoked. When the court dismissed his petition, Quelch filed an appeal, which was subsequently settled out of court.[9] *The Times*, which followed the case, reported that advertisers eagerly paid to put their shops on the map.

More is known about Robert Gatliff, who was a "window trimmer" at the Railey-Miram Hardware Company in Miami in 1921, when he filed a patent application for a directory map similar to Merk-Wirz's.[10] Born in Kentucky in 1890, he had worked in retail hardware as a salesman and was building a reputation as a window dresser.[11] According to newspaper articles from the 1930s and 1940s, he also crafted ladies' hats and dresses from steel wool, paint brushes, wire mesh, tea kettles, and other stock items—absurdly ostentatious gimmicks for drawing customers into the store.[12] Gatliff worked on window displays into his 60s, much of the time at the same firm, and in his early 30s knew enough about bulbs and wiring hardware to design an innovative electric map.

Gatliff's main drawing shows multiple destinations listed by name (12 in Fig. 7.2) in the margins of a large-scale map titled "Automatic City Directory." Next to each name is a pushbutton (13) linked electrically to a corresponding lamp on the map (11) as well as to the single you-are-here lamp (14). As with Merk-Wirz's invention, pressing the button momentarily connects both lamps in series with a battery (described on

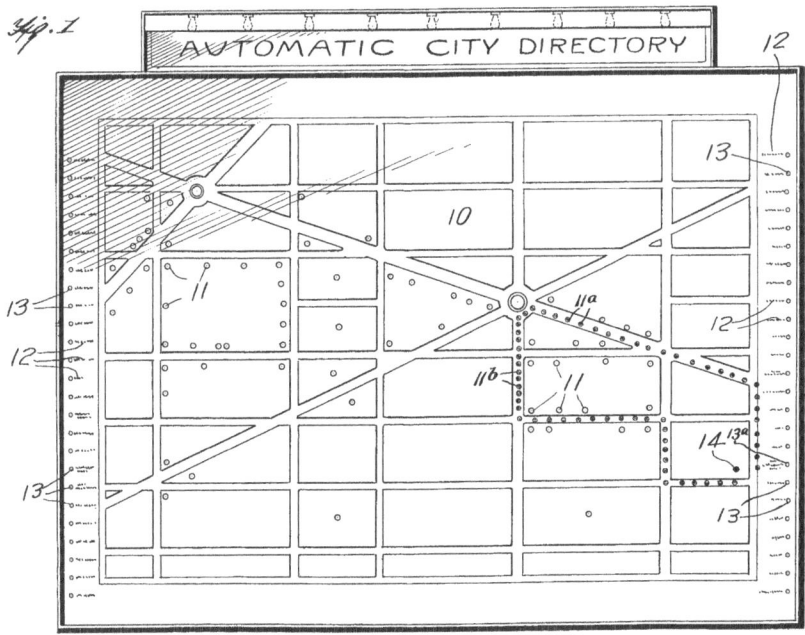

Fig. 7.2 Robert Gatliff's patent for an electric map added alternative routes to a selected destination (US Patent 1,409,894; 1922)

another drawing). What's innovative is two optional routes, each marked by a string of bulbs (11a and 11b) embedded in the street symbol, with different colors indicating travel by foot or streetcar. Gatliff is silent about the complex wiring needed to describe routes to multiple destinations.

Gatliff's application encountered moderate resistance from a patent examiner who pointed out similarities to Merk-Wirz's invention.[13] But nine months later, after three letters of rejection and a reduction from four claims to one, the patent was issued. Half the rights were assigned to Robert R. Reimert, Jr., a local architect, who might have offered to help with development and marketing. In July 1929, Gatliff filed another patent application, for an "Advertising Device" that enhanced a billboard with electrically driven characters, including a man who raised a Coca-Cola bottle to his mouth and tilted his head back as he drank.[14] This was Gatliff's second and last patent, and I found no evidence either his animated sign or his markedly more complex directory board was ever licensed or manufactured.

The next patent in my list of electrical maps was definitely developed, though probably not as fully as its inventor hoped. In August 1930, the same month he filed his application, 25-year-old Edward R. Swett, Jr., sold an "Electric Fire Alarm Map" to the city of Grand Rapids, Michigan, for $4960.50 (about $70,000 today). As described in the *Proceedings of the City Commission*, the system would "enable the Fire Department to expedite and lessen the chances of error in recording and answering fire alarms."[15] A promotional booklet by "Edward R. Swett, Jr. and Associates, Fire Alarm System Engineers" included photos of installations in both Grand Rapids and Muskegon, Michigan, a smaller city 40 miles away, where the firm was located.[16] The booklet accompanied a letter to the Patent Office from Swett's attorney, who claimed, "the system is in use in a number of different citics and is meeting with great success."[17]

Swett's invention juxtaposed an electric street map with a switchboard listing the names of all city streets, large buildings, and other well-known places (Fig. 7.3).[18] A drawing for one of the zones (Fig. 7.4, left) showed each street's line symbol enhanced by at least one "indicating light" (3), illuminated by toggling the switch (7) next to the street name. Names were arranged alphabetically, so that a telephone operator at the central fire station could determine the street of the fire, turn on the corresponding light, ask for the name of the nearest cross street, and illuminate a second light. For example, if a fire were reported near the intersection of Apple and Wood streets, the lights for this pair of streets would place

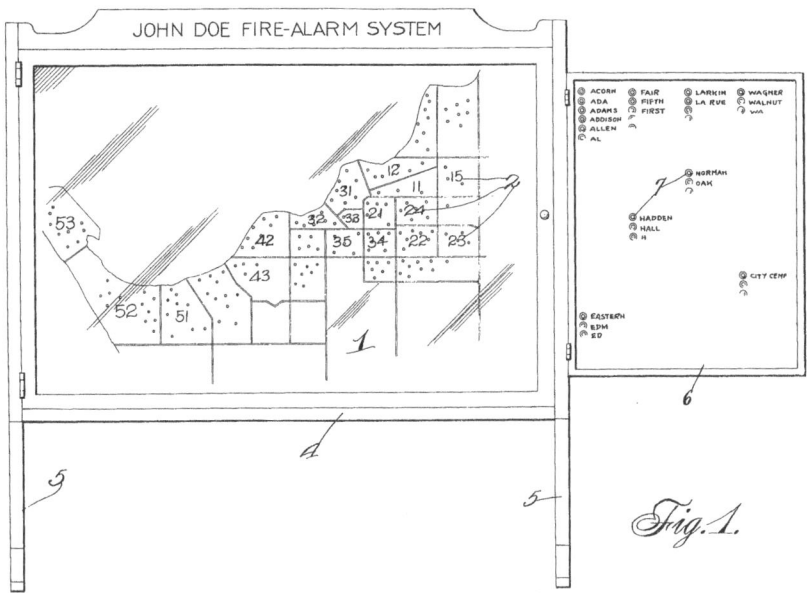

Fig. 7.3 Edward Swett's fire alarm system linked an electric street map (*left*) to a switchboard with street names (*right*). The map divided the city into zones marked with large numbers (11) (US Patent 1,929,759; 1930)

the fire near the southeast corner of block 15 in zone 11. As with a signal telegraphed from an alarm box, the fire's location was only approximate.

Another drawing (Fig. 7.4, right) described an index card with a planned response to an escalating fire near this intersection. The "11–15" on the tab and top line of the card was a four-digit code sent by wire to every "engine house, truck house, and hook-and-ladder company" in the city. In this example, a gong would ring once, and after a short interval once again, to indicate zone 11. At the appropriate firehouses, fire fighters would then listen attentively to the next two digits (a single gong, followed by five gongs in a row), indicating block 15. The corresponding card indicated the squads and apparatus expected to respond. Each firehouse was given a full set of these cards as part of a comprehensive system that included delineating zone boundaries and assigning block numbers.

Swett's invention was a cartographic enhancement to existing fire communications, whether telephone or telegraph. As long as the telephone

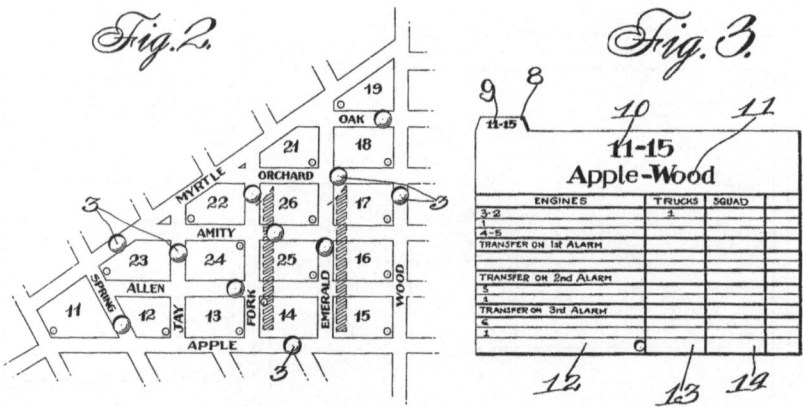

Fig. 7.4 Separate drawings described a hypothetical zone (*left*) and an index card relating the street intersection to programmed responses for first, second, and third alarms (*right*) (US Patent 1,929,759; 1930)

operator at the central station understood the street names correctly, there was little likelihood of a mistake in writing down the location or plotting it on a map. Indeed, if the caller reported two streets that did not intersect, the operator knew to ask for clarification. In this sense, Swett's system gave telephoned reports the reliability of a network of fire telegraph alarm boxes, which transmitted the box's exact location to a central station as a string of numbers. Expansion of residential telephone service, and the consequent ease of reporting a fire, helped promote Swett's invention.

However marketable, Swett's system did not move swiftly through the Patent Office, which rejected his application three times and reduced the number of claims from 11 to 1.[19] The case file shows polite wrangling over wording as well as possible interference with Merk-Wirz's patent, among others. The examiner also spotted a cartographic glitch Swett and his attorneys were apparently allowed to ignore, namely, that "Fig. 2 (Fig. 7.4, left) shows two lines of oblique-line areas adjacent to portions of two of the streets but no explanation of these is found."[20] These blatantly extraneous symbols could easily have been erased. Either the glitch was overlooked in the final examination or patent officials felt Swett and his attorneys had ceded enough ground.

Though Swett's fire alarm business was apparently never more than a sideline, it probably helped him start a new career. The 1930 Census found him working as an assistant manager in the hotel industry, probably at Muskegon's Occidental Hotel, the address of his fire alarm business—the Occidental was the city's premier hotel, and his father was the general manager. For the 1940 Census, which asked about the highest level of education, he reported two years of college, employment as a "commercial representative" in the telegraph industry, a solidly middle-class annual income of $3900, and an equally middle-class family with a wife and three children. Swett never applied for another patent and at some point retired to Florida, where he died in 1981 at age 76. Aside from a possible link to telecommunications in the hotel, his sketchy biography offers little clue to how he conceived his electric map.

Oddly, the Gamewell Company, the country's most prominent manufacturer of fire telegraph equipment, made no provision for identifying its alarm boxes on a map.[21] Because my Printed maps dataset might have missed other relevant inventions, I searched Google for patents filed before 1931 with the terms *map, box, fire alarm,* and either *electric* or *circuit* anywhere in their text. The search found five patents, but only one included an electric map similar to Swett's.[22]

That patent had been issued in 1918, more than a decade before Swett's, to Nathaniel Banks Cregier, an electrical engineer in Chicago. An accomplished inventor, Cregier had received 11 patents between 1891 and 1932 for inventions as varied as a record player, an automatic violin, and a safe with an electrical time lock.[23] His "Signal System," which connected fire alarm boxes throughout the city to a central station, was arguably more sophisticated than Swett's. A citizen who pulled the lever at an alarm box sent an electrical impulse that rang a gong at fire headquarters, where the box's location was registered automatically on a recording mechanism and displayed on a street map. The patent described the configuration of wiring circuits, electrical relay switches, bells, batteries, and lamps with seven diagrams, one depicting part of a larger map with street names and lamps (50 and 51 in Fig. 7.5) indicating box locations.

A similar map excerpt was embedded in the circuit diagram (Fig. 7.6) for the police "Signaling System" Cregier had patented in 1909.[24] Unlike his pull-the-alarm, one-way fire alarm system, this earlier system linked a central or divisional police station to two-way call boxes with telephones. Another diagram showed the street map as a key element in a central switchboard, with which a call box could be connected to another telephone inside the police station (Fig. 7.7). As with Swett's system, intersecting streets served as coordinates for locating incidents.

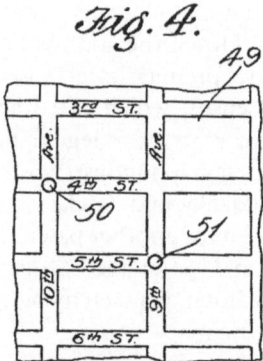

Fig. 7.5 Portion of the street map display integral to Nathaniel Banks Cregier's fire alarm system (US Patent 1,274,514; 1918)

Neither invention was readily endorsed by the Patent Office. The police alarm encountered four rejections and was not patented until exactly five years after filing. The fire alarm was even more problematic, with five rejections and more than seven and a half years of intermittent and sometimes contentious correspondence.[25] In the first volley, an examiner recommended dividing the application into three sets of claims—for a telephone system, a fire telegraph, and a composite "fire or police" telephony system—each to be vetted in a different section.[26] Cregier's attorneys, a Los Angeles law firm, responded by rewording 5 of the original 29 claims, adding a 30th, and asserting that "the requirement of division has been obviated since the features of the system which make it distinctly a fire alarm system have been included in all the claims."[27] Three months later, an examiner rejected all 30 claims, mostly because of possible preemption by an existing patent.[28] After further wrangling, the examiner rejected 10 of the 27 claims still on the table, and concluded, with polite exasperation:

> This application has now been pending for some time and the prosecution thereof has extended through a large number of actions, and as no new references have been cited and no new points have been raised, the rejection of the claims in this action is hereby made final. An amendment which will enable the examiner to pass the case to issue will be admitted after final rejection.[29]

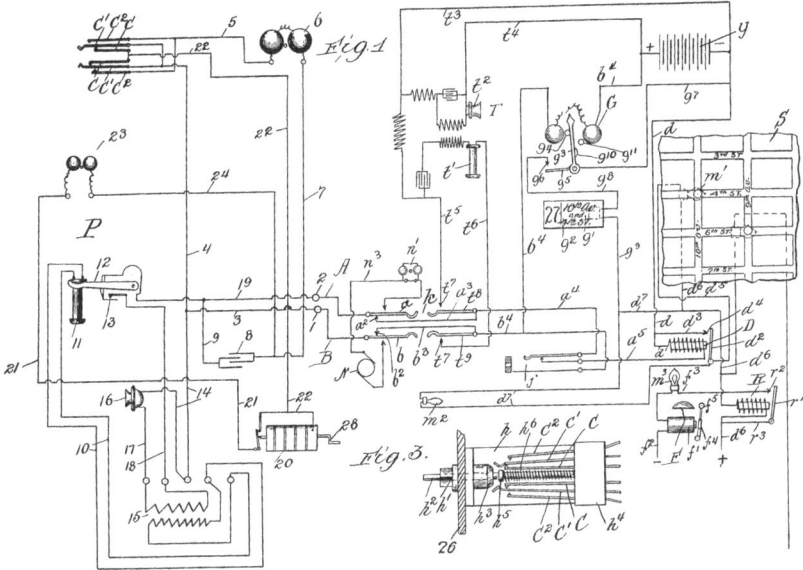

Fig. 7.6 Circuit diagram for Nathaniel Banks Cregier's earlier police signaling system. Note the street map on the right side, just above center (US Patent 915,075; 1909)

Although the examiner's tone suggests the Patent Office had had the final word, Cregier's attorneys argued for retaining four of the ten rejected claims.[30] Although the examiner addressed these latest arguments, he rejected the amendment because "the case is under final rejection and the amendment is not of such a nature as to place the case in condition for allowance."[31] The attorneys then complained that the examiner had misnamed a reference ("Goldstein, et al, not [merely] Goldstein") and contended that, "In view of the fact that a great number of references have been cited it is apparent there would be a great waste of time on the part of the attorney in searching out the references to find out which one the Examiner had meant when it was apparent that an error had been made in naming the reference."[32] One claim had been cancelled and others amended, they noted, and "the case is now in condition for allowance." But only if the attorneys agreed with the examiner, who sent a copy of the regulation giving the Patent Office control over final edits along with a

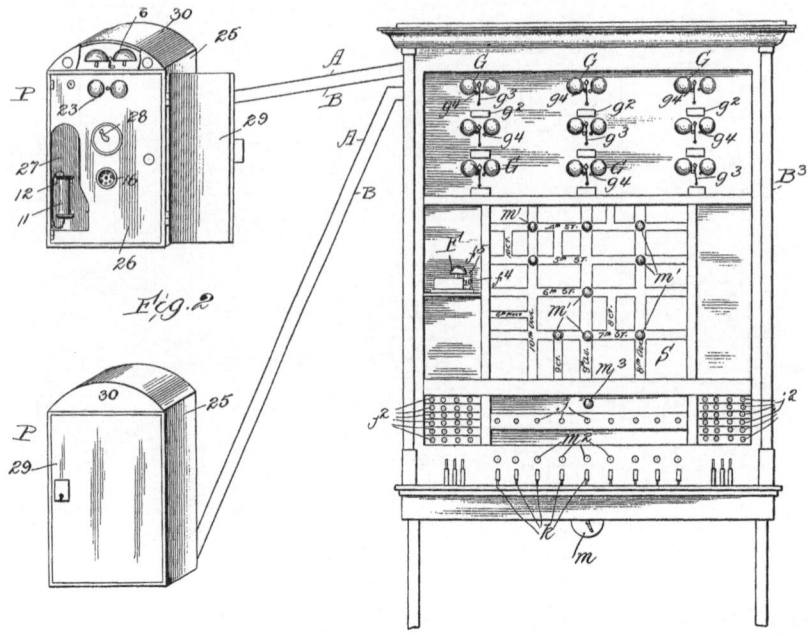

Fig. 7.7 Front view of Cregier's central switchboard, juxtaposed with interior and exterior views of police signal boxes. Note the bell inside the signal box (*upper left*), used to summon a nearby police officer (US Patent 915,075; 1909)

formal notice that the patent had been allowed and would be issued after payment of the "final fee."[33] Pay the fee, accept our changes, and have your patent.

Cregier's attorneys apparently never informed the inventor. Nearly two years lapsed before the Patent Office sent him a telegram—collect, on Christmas eve—advising that his application, which had been allowed but still required final payment, would be abandoned unless renewed in 17 days.[34] He hurriedly found a Chicago lawyer, who mailed a new power of attorney to the Patent Office, along with a $15 check to renew the application.[35] Seven months later, he had his patent, now with only 11 claims.

Hiring a Los Angeles law firm had made sense in 1911, when Cregier applied for his patent. Three years earlier, he had incorporated the Cregier Signal Company, in Chicago, to produce and market fire and police alarm systems,[36] and in 1909 he was in Los Angeles vying for a city contract at what the *Los Angeles Herald* called "one of the most distinctive and novel exhibitions ever held in the United States."[37] Six alarm makers competing for a $100,000 city contract demonstrated their systems to city officials and the public in a vacant department store. The competition included Gamewell and the German firm Siemens & Halske.

Although Cregier was not the low bidder, he won a contract for 100 fire alarm boxes, installed in the University district the following year.[38] An article in *Insurance Engineering* that described the system's electronics, operation, and placement of alarm boxes quoted extensively from a report by the Committee on Fire Prevention of the National Board of Fire Underwriters, which noted a variety of deficiencies, including high installation cost, but pronounced the system "well designed for the district protected."[39] The report noted, "The illuminated map is an attractive and somewhat useful feature, but is no essential part of the system."[40]

Halfhearted enthusiasm for Cregier's electric map is understandable. Central fire stations always included a large-scale city wall map near the alarm board, and experienced operators were familiar with the local street grid.[41] Even so, an automatic map might save vital seconds. According to Oscar Levy, an electrical engineer who patented a more sophisticated fire alarm system in the early 1930s,

> ...it may be desirable to arrange all the lamps associated with the corresponding fire alarm boxes upon a map of the community or town, thereby showing instantly not only the number of the calling fire alarm box but also its location and the location of the nearest fire station. This is an advantageous arrangement, as it eliminates the possibility of error due to reading a signal of the lamp associated with the calling fire alarm station and then consulting the record or cards for its location in the system, thereby saving considerable time in sending out the fire alarm apparatus to the fire.[42]

Even so, this is the only instance of *map* in Levy's patent, which did not include a map in any of its drawings. Nowadays, a zoomable cartographic display is an obvious and essential component of any electronic emergency response system, but a century ago it was an expensive add-on.

How did Cregier learn about electricity and fire alarms? Although he apparently lacked a college degree or military training, Banks Cregier (as he preferred to be known), was the third son of DeWitt Clinton Cregier, who served Chicago as city engineer before a two-year term (1889–91) as mayor.[43] Born in 1863, Banks worked for the city from 1884 to 1891, when he received his first patent, for a "Police Signal System" more rudimentary than the map-enhanced version patented nearly two decades later.[44] He assigned half the rights to his father, whose experience with patents included four of his own, three for fire hydrants.[45] The son's fire alarm patent languished for several years, until the newly emerged Stromberg-Carlson Telephone Company was able to manufacture the apparatus.[46] His interest in fire alarms might have been inspired by the Great Chicago Fire of 1871, which had destroyed a third of the city as well as the family home.

It is not clear how many cities bought Cregier's map-enhanced fire alarm system.[47] He apparently worked as an engineering consultant to various municipalities, including Chicago, which hired him in 1927 to assess its fire and police telegraph systems. The 1930 Census found him working as an electrical engineer for the city fire department, and a family history lists him as "supervising engineer" of the fire department "until his death" on 10 January 1935, at age 71.[48] At some point he had sued the city for $135,000 for infringing his 1918 patent, but the case was settled in the late 1930s, probably by his heirs, for a mere $5000.[49]

Despite fundamental differences in the electronics of one-way telegraph and two-way telephone alarm systems, patent drawings like Swett's and Cregier's clearly highlight the electronic map's role in quickly showing an incident's location, however imprecise. By contrast, the "Educational Apparatus" patented in 1933 by the Pollard brothers, Robert and Oscar, of Indianapolis, is less straightforward, thanks to awkward mechanical adjustments, non-intuitive map symbols, and clumsy patentese.[50] The device is obviously a map (Fig. 7.8), intended to promote geographic education by quizzing pupils about place names, but the time required to set up the questions seems excessive.

Understanding the Pollards' invention is easy if you treat the map as a quiz in which place names toward the upper left are the questions, locations on the map are the answers, and a light bulb flashes when a pupil gets the right answer. Along the left edge a vertical conductive rod (6 in Fig. 7.8) holds a sliding conductive pointer (8) with a setscrew (3) and light bulb (13). The teacher begins by positioning the pointer near a place name such as Montevideo, carefully aligning the point (9) with an electrical contact represented by a small circle with a dot in the center (4), and tightening the setscrew. A concealed wire running underneath the map's nonconductive surface links this contact with a contact at the correct loca-

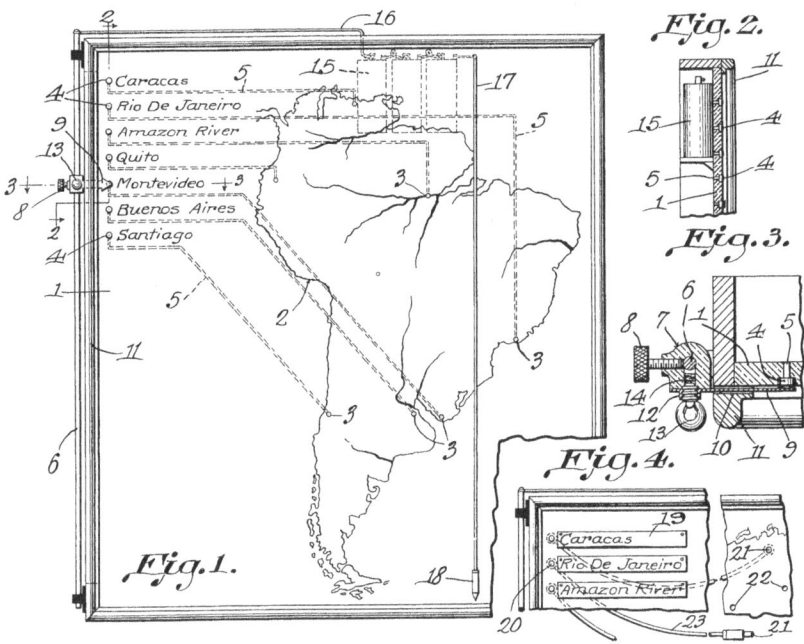

Fig. 7.8 Drawings for the Pollards' "Educational Apparatus" (US Patent 1,932,909; 1933)

tion on the map, represented by a small circle (3). A knowledgeable pupil completes the circuit by touching this second contact with the conductive pencil-like pointer (18) at the end of a conductive cable (17) linked in turn to a low-voltage battery pack (15) and a wire (16) running across the top of the map to the vertical rod. The Pollards' narrative obscures the process by calling the flexible cable (17) a "rod" and portraying it as straight and rigid—an inflexible arrangement that just won't work.

Other options are more promising. A buzzer can replace the light bulb, and although one teacher can handle multiple pupils simultaneously, a patient student fascinated with electricity could play the game alone. Although the Pollards framed their invention around geographic instruction, they recognized that its electric feedback scheme could be adapted to rote learning in "mathematics, languages and other studies."

It was hardly surprising to discover that both Pollards had been teachers. When they filed their patent in July 1932, Robert was 37 years old and Oscar was 35. Records of the 1930 Census list both Pollards as public school teachers, but a 1932 city directory reported Oscar working at the Post Office, and draft board records indicate that by 1942 both were postal workers. I found no record of any prior or subsequent patent. Their experience with the Patent Office was equally unremarkable: a single rejection reduced four claims to one.[51]

Like most patented inventions, theirs was never developed commercially. The idea was clever, but where classroom engagement did not demand gadgetry, even a neophyte teacher could get similar results less expensively with a simple outline map. Even so, their patent was referenced decades later by examiners vetting four patents, including one titled "Educational Device for Learning Geographical Names and Locations."[52] Approved in 1984, the latter patent referred to a map board with multiple lamps, one for each state, but hid the electronics on cartridges described only by function, not by circuitry, thereby circumventing an abundance of complex wiring diagrams and needlessly clunky wording.[53] Inventors of electronic equipment—and in later years computer software—benefitted enormously from the Patent Act of 1952, which allowed claims to be described largely by function, rather than by their operational details.[54]

The last two patents in my core group of six reflect a transition from analog to digital cartography, that is, from maps created from physical materials like paper, ink, plaster, plastic, cloth, light bulbs, and simple electronic hardware to maps represented by numbers manipulated by a digital computer. The earliest of the two is a terrain map consisting of a thin copper sheet resting on a non-conductive base and divided into zones by thin bands of non-conducting material that represent elevation contours. The patent drawing in Fig. 7.9 is a plan view of a map in which each zone represents a 20-foot range of elevation, and the number inside the zone is the elevation of the lowest of the bounding contour lines. For example, the contour labeled 14A near the center of the map has an elevation of 580 feet and separates the 580–600 zone (labeled 580) from the 560–580 zone (labeled 560).

Note that wires connect all zones for the same elevation interval, as illustrated by the three zones labeled 620 at the top of the drawing. Also note the two zones labeled 580 and bounded by closed contours lines in the lower half of the drawing; dashed lines representing wires below the insulated base connect both of these zones to the larger zone labeled 580 that winds from right to left across the map. Similarly, another dashed line in the lower-right quarter of the map depicts a hidden electrical connection between an isolated 560 zone and a larger region for the same elevation interval.

A wiring diagram along the left side of the drawing shows how electrical resistance represents elevation when the "scanner contact" (36) at the

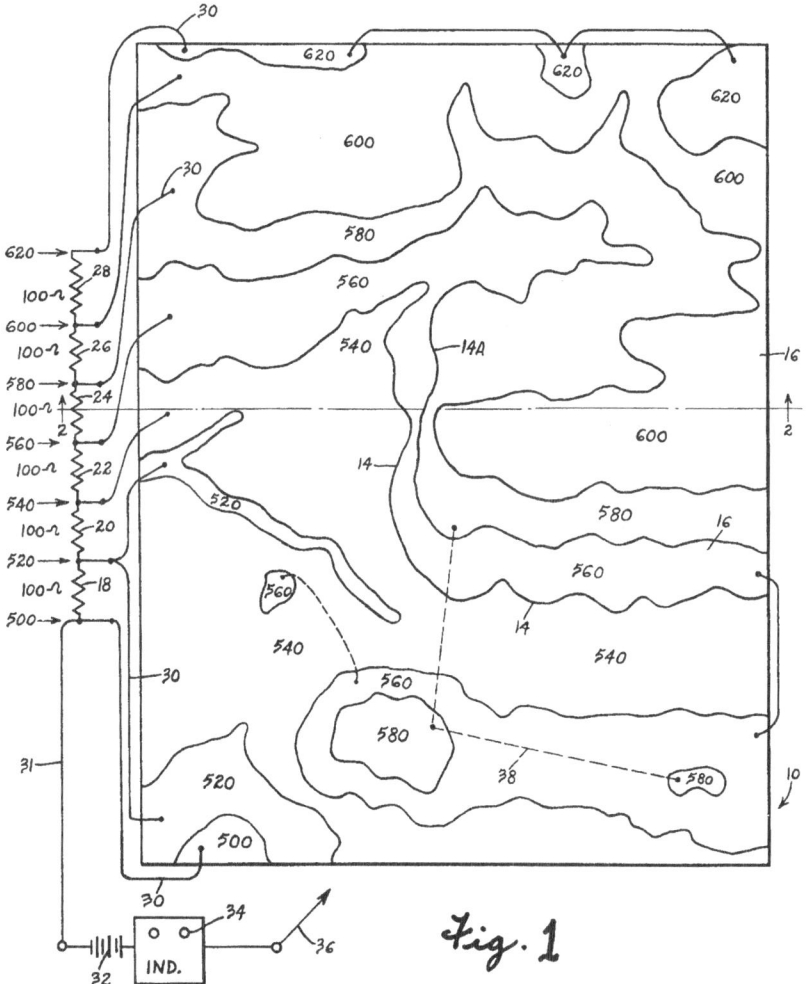

Fig. 7.9 Joseph Stieber proposed an electric map on which non-conductive contour lines (14) etched into a thin copper sheet partition the area into 20-foot elevation zones (US Patent 2,876,562; 1959)

bottom of the drawing is extended to touch one of the zones. Touching the small 500 zone near the lower-left corner of the map will complete a circuit that includes a battery cell (32) and an "indicator" (34) for measuring current flow. Current flow is measurably less when the contact touches one of the map's two 520 zones because the circuit now includes a 100-ohm resistor. The wiring diagram shows the addition of another 100-ohm resistor for each successive elevation interval, so that when the scanner contact touches one of the 580 zones, for instance, the indicator reflects a combined resistance of 400 ohms.

This electrically readable elevation map is the analog equivalent of what cartographers call a digital elevation model (DEM), whereby elevation measurements are organized in rows and columns like the cells on a sheet of graph paper. Moreover, this electric analog map could be converted to a digital map by a scanner that samples the map systematically at points representing cells on an imaginary sheet of graph paper placed over the map, converts each measured current to a number, and stores these measurements in a DEM. To capture the intricate details of a typical analog map requires a scanning increment much smaller than the resolution of a typical sheet of graph paper. Once digitized, the map could be used to display a three-dimensional view of the terrain or drive the creation of molded plastic relief models like those produced by the American military after World War II.[55]

A link to military terrain mapping is implicit in the patent's lead sentence, which notes that the invention "may be manufactured and used by or for the Government of the United States for governmental purposes without the payment of any royalties"—understandable insofar as the inventor, Joseph A. Stieber, was a government employee, represented by attorneys at the Office of Naval Research (ONR).[56] No Patent Office fees were required, but an appropriately thorough examination yielded three rejections, which reduced the 11 original claims to 7. The patent was granted in 1959, exactly four years after the date of filing.

For whatever reason, the examiner also tinkered with the invention's title, which was changed abruptly on the Notice of Allowance from "Potential Levels for Automatic Scanning of Contour Maps" to "Electrical Method and Means for Making Relief Maps."[57] The case file at the National Archives contains no explanation—just the annotation "AS AMENDED BY EXAMINER"—and no objection from Stieber's attorneys. This late-in-the-game substitution is puzzling because *potential* aptly refers to the role of *electrical potential*, or voltage, in producing a measurable current.

That said, *method* and *means* echo key goals of the invention, namely, "a simple method to electrically scan a map and convey that information to a computer for conversion into a master model of the terrain" and a means "to avoid the necessity for skilled technicians." If not a prophet, Stieber was clearly anticipating the Third Industrial Revolution: although humans could carry out data conversion, they not only required training but were slower and more error-prone than computers.

Why did the ONR sponsor Stieber's patent without retaining all property rights? The simple answer is a wariness of infringement lawsuits, which had plagued the Navy after World War I.[58] Although few damage claims were actually paid, an effective defense could be expensive and distracting. Sponsoring patents like Stieber's not only allowed the government to use the invention royalty free but also reduced the likelihood of infringement claims by other inventors. In effect, Stieber's application was a partnership that credited him with a clever idea and further shielded the Navy against lawsuits. That the inventor was free to license the patent in the private sector was inconsequential.

Was Stieber's invention ever employed for map production, by the military or anyone else? Probably not. The contour lines would have been etched into the copper plate by a photomechanical process, and preparing an existing map for photoengraving would have been a tedious, labor-intensive step, as would the wiring of a large, cumbersome electric map with a multitude of resistors. What's more, electromechanical drum scanners with a suitably high image resolution emerged in the early 1960s to provide more direct analog-to-digital conversion of traditional contour maps.[59]

For an inventor who helped father digital cartography—along with many others—Joseph Stieber left a surprisingly faint biographical footprint. I could find no evidence of higher education, military service, or even place of employment. His patent describes him as a resident of Valley Stream, New York, a workable commute from the ONR's Special Devices Center, a research laboratory in Port Washington, on Long Island's North Shore, but I found no direct connection.[60] I did find several people named Joseph A. Stieber, the most likely of whom was born in 1910, resided in New York when he applied for a Social Security card, and died in Fort Myers, Florida, in 2006, at age 96, without an obituary. Three years before his death, this Stieber married a 76-year-old widow, who lived to age 85—a fact worth noting only to confirm my having searched.[61]

Joseph Stieber's cartographic legacy includes two other patents, both filed in early January 1956, the year after he applied for the electric map patent.[62] Both patents were teaching aids for instruction in map projections, both gave the government royalty-free use, and both were collaborations with an equally obscure Levittown, Long Island, resident, John B. Weldon. Titled "Map Projections Demonstrator" and "Methods for Forming a Color Impregnation of Transparent Geometrical Shapes," the inventions are training-oriented R&D projects in line with work at the ONR's Long Island facility.

ONR attorneys also sponsored the remaining patent in my core group: the "Electrically Coded Terrain Model Map" invented by Edward G. Valliere, of Elkins Park, Pennsylvania, a suburb of Philadelphia, and filed in November 1958.[63] Unlike Stieber, Valliere is not identified in the patent's case file as a government employee. That he assigned all rights "to the United States of America as represented by the Secretary of the Navy" suggests he might have worked for a Navy contractor, which could have been American Electronic Laboratories (AEL), in nearby Colmar, Pennsylvania—a rough guess based on his having signed over to AEL non-cartographic patents he had filed in 1962 and 1975.[64]

Like Stieber, Valliere left a weak biographical footprint, muddled by multiple people with the same first and last names. The only reliable connection is his graduation from Philadelphia's La Salle College (now La Salle University) in 1965 with a B.S. in Electronic Physics, a six-year program in the Evening Division. His yearbook entry reports an address in Roslyn, Pennsylvania, another Philadelphia suburb, and he looks more than a few years older than most of his classmates.[65] His immediate government liaison could have been located at the Philadelphia Navy Yard, which specialized at the time in servicing non-nuclear warships with comparatively complex electronics.[66]

Valliere's invention was part of a larger endeavor that included Stieber's electric contour map and four other patents identified by number in the second paragraph. All four were filed on or within two weeks after his November 20 filing date. While the patent for an "Automatic Wave Analyzer," assigned to General Dynamics Corporation, in San Diego, has no apparent link to mapping, the other three are clearly cartographic. Two focused on map production—the "Coordinate Positioner" for reproducing contours on a plotting table, and the "'Z' Axis Drive System" for raising, lowering, and rotating a cutting tool for making a three-dimensional relief model—and the third, fittingly titled "Contour Data Recording System," could store data captured by scanning Stieber's electric map.[67]

All three map-related patents were filed by Philadelphia-area inventors and assigned to the Secretary of the Navy.

A single page of text describes Valliere's invention, intended as an efficient method for "making an accurate, low cost electrically coded metal foil contour map." He preferred copper foil as a reliable electrical conductor that can be etched photographically and then coated with tin to promote soldering and resist oxidation. A single drawing describes six components, identified by both numbers and labels and arranged in five layers (Fig. 7.10). Second from the top is a layer of metal foil (10) that has been partitioned by photoetching into zones connected by wires to other zones representing the same elevation interval. Each zone is also connected electrically to a rectangular terminal representing its elevation interval; for convenient access, all elevation terminals are arranged in a row (18) along the near edge of the map. A layer of vinyl adhesive (14) anchors the foil (10 and 18) to a temporary metal backing (12) with a vinyl protective layer (16) on the bottom. After the wires are soldered in place, a non-conductive casting resin (24) is poured on top and allowed to fill the gaps (contour lines) between zones. The resin hardens to electrically isolate each zone from its neighbors. Peeling away the lower three layers and inverting what's left yields a laminated embodiment of Stieber's electric

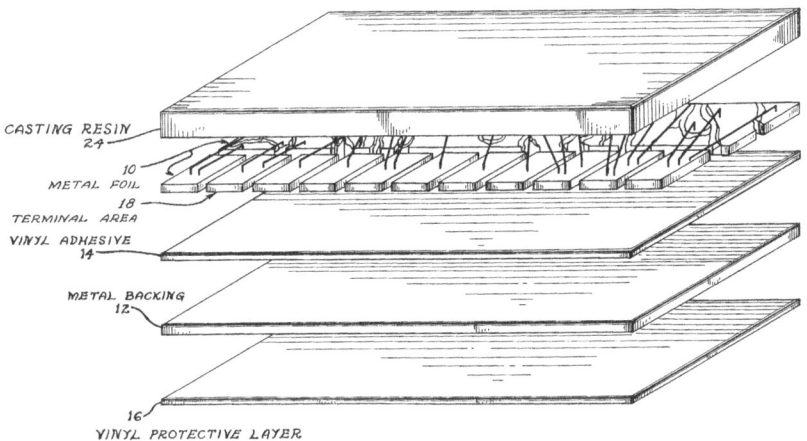

Fig. 7.10 Edward Valliere identified his invention's five layers with both labels and numbers (US Patent 3,097,418; 1963)

map, with the foil exposed and good to go once resistors are applied to the elevation terminals.

Valliere's fabrication method endured a prolonged dialectic dance with the patent examiner, Whitmore A. Wiltz, who pared the application's eight original claims pared down to three. In the first of four letters of rejection, Wiltz declared all eight claims "indefinite."[68] Valliere's attorney, Lawrence Epstein, responded by cancelling five of the claims, changing a few words, and adding three new claims, which Wiltz then rejected.[69] A claim focused on the encapsulation of foil zones in resin was particularly contentious because Wiltz considered the process covered by an existing patent and thus not patentable.[70] In rebuttal, Epstein denounced the examiner's opinion as "a self-serving statement [that] should be supported by prior references in order to be effective."[71] Unimpressed, Wiltz insisted the process "would not require invention" and concluded, "This action is made FINAL."[72] Unwilling to capitulate, Epstein complained to the Patent Office's Board of Appeals, which apparently agreed because the disputed claim emerged largely intact when the patent was issued in July 1963, more than four and a half years after filing.[73]

Although Valliere's invention most likely was never implemented, his patent (and those for related inventions) heralded the numerical cartography that came to dominate mapmaking and map use in the late twentieth century. Instead of electrical circuits for activating light bulbs or current meters, the digital computer empowered software-generated maps as well as an explosion of interactive, online, and mobile mapping applications.[74] This transition is apparent in time-series plots created for *electric map* and *digital map* (Fig. 7.11) using Google's Ngram Viewer, which tracks words and phrases in the millions of books scanned for the Google Books Project.[75] The horizontal axis covers the period 1910 to 2008, and the vertical coordinate represents the term's relative share of all two-word phrases in books published each year. *Electric map* attained its greatest relative prominence in the 1930s, but was eclipsed by *digital map* in the mid-1970s.[76] Both terms gained ground in the 1980s but declined more recently as numerous new cartographic forms, like GPS map, Google Map, and weather radar map, entered everyday usage. *Electronic map* (not shown), which emerged in the early 1940s and overtook both terms from the late 1960s through the mid-1970s, experienced a similar rise and fall, to finish the period with only a quarter of the prominence of *digital map*.

In the digital era, map-related patenting shifted from the independent inventor to the corporate sector, where patents became assets as well as

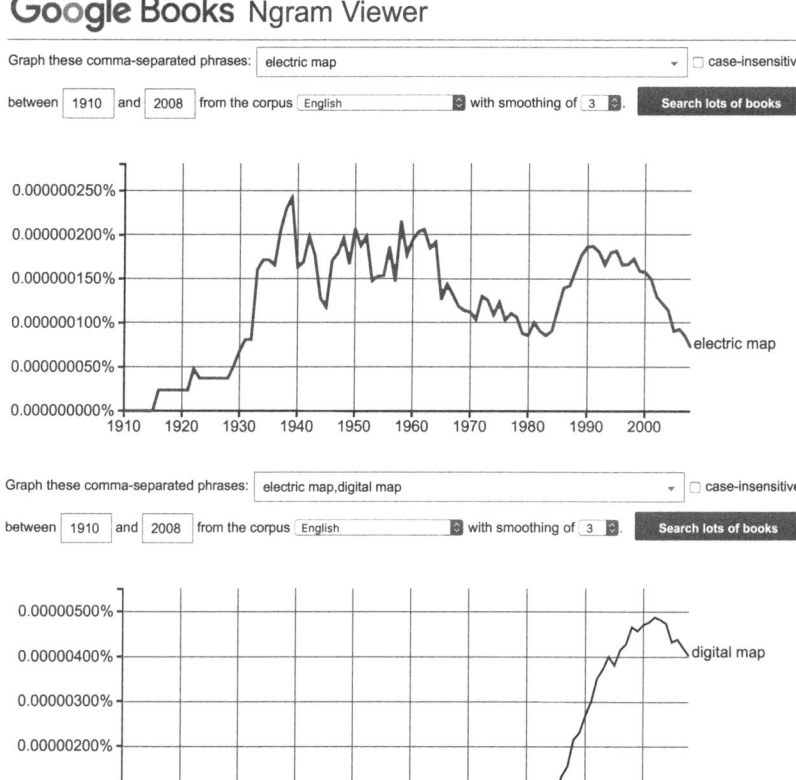

Fig. 7.11 Google Ngram Viewer time-series reports for *electric map* and *digital map*. Formatted by the author

defensive weapons against infringement claims. The individual innovator's need for achievement, satisfied in earlier decades by one or two patents (most never developed commercially) survived more or less intact when corporate employers began stockpiling hundreds of patents, which often listed multiple inventors. In this milieu, getting one's name on a patent, no matter the number of co-inventors, became a tangible measure of success—the firm owned the rights and the inventors got the credit. Recent examples of multi-inventor patents include Apple's "Mapping Application with Interactive Dynamic Scale and Smart Zoom," which lists seven inventors, and Microsoft's "Reduced Power Location Determinations for Detecting Geo-fences," which lists nine.[77] Although corporate patents often name only one or two inventors, multiple authorship is an inevitable consequence of the intellectual property arms race.

Where does this leave the patent system as a parallel literature comparable to academia's refereed journals? Look no further than high-cost international endeavors like high-energy physics, where a coalition of scientists connected with the Large Hadron Collider recently published a paper with 5154 authors.[78] Although collaborations this massive are rare, multi-authored cartographic articles, like their counterparts in the patent system, have become more common in recent decades.

A new patenting phenomenon without parallel in academia is the patent troll, also called a "non-producing entity" or "patent-assertion entity" if you prefer a less pejorative term.[79] In the troll's world, a patent is a license to litigate rather than merely a legal right to make and sell a product. A troll can acquire patents through invention or purchase, and some law firms specialize in representing trolls in infringement lawsuits, ideally with multiple defendants—the more the better when the goal is to extract (dare I say *extort*) an out-of-court settlement from each defendant.[80] Although settling a lawsuit is seldom cheap, it is immensely more reliable than letting the case go to trial and typically less expensive than a strong defense: good attorneys and convincing expert witnesses can be costly, as the Navy discovered after World War I. Moreover, a settlement is almost always less costly than whatever damages might be awarded by jurors who resent a corporation's wealth and see the lawsuit as a *David v. Goliath* contest.

Infringement litigation has created a market for overly broad patents, which patent trolls buy and sell like commodities. Perhaps the most flagrant case involving a map-related patent originated in 1996, when three Southern California residents filed an application titled "Internet Organizer for Accessing Geographically and Topically Based Information."[81] Issued in 1999, their patent was traded several times before the Antigua-based holding company Ubixo Limited bought it for $119 million and transferred the asset to its Texas-based spin-off GeoTag Inc., which in mid-2010 started filing infringement actions against more than 300 companies, including Wal-Mart, Home Depot, and Lowes.[82] Ten separate lawsuits, each with multiple defendants, were filed in East Texas, where the Federal District Court for the Eastern District Texas (and its juries) has been the scourge of high-tech firms sued for patent infringement.[83]

GeoTag's patent describes a database retrieval system for helping users find information about goods and services available either locally or across the region. Information is stored in electronic folders organized by topic, like a yellow pages telephone directory, or hierarchically by place. The software interface includes an interactive "viewpoint" map customized for particular areas or regions, and 39 interlocking claims describe the system's organization and capabilities. Focused on functionality, rather than on operational details, the claims had made it easy to argue for infringement by nearly any online system designed to help potential customers find a nearby store and probe its inventory.

GeoTag's targets typically relied on largely generic systems powered by software and map data from Google and Microsoft, which in early 2011 jointly filed a lawsuit to invalidate GeoTag's patent.[84] Their strategy was to request a "summary judgement," whereby a judge assessed the merits of the case without a full trial before a jury. The District Court left the patent intact but ruled against GeoTag because infringement does not occur when demonstrably different methods yield the same result: a conclusion the multiple defendants had been reluctant to test in court. GeoTag appealed and the Circuit Court upheld the lower court's decision.[85]

My interest in patent trolls began in summer 1999, when a law firm representing Garmin International hired me to review materials in *Civix v. Microsoft, et al.*, filed six months earlier in the Federal District Court for Colorado.[86] Garmin, a well-known manufacturer of handheld GPS and vehicle navigation systems, was part of the "et al.," along with Magellan (another GPS navigation firm), Rand McNally, Ticketmaster, Yahoo!, America Online (AOL), Bellsouth, the Denver Post, and 11 other firms alleged to have infringed one or more claims in two patents, issued in 1990 and 1997.[87] According to the filing, "Both patents generally relate to electronic mapping systems which can be used to locate businesses and other points of interest in a defined region." The nine-page complaint demanded a jury trial.

I flew down to New York and spent a day with the three attorneys working on Garmin's case, including a corporate lawyer from company headquarters, near Kansas City. I was astonished that Civix's complaint hinged on technology well known in the academic community in 1990 and 1995, when the patents were filed. In my opinion, the patents, though codified by 7 and 37 claims, respectively, hardly met contemporary requirements for novelty and non-obviousness. In particular, the 1990 patent describes a system no more functionally sophisticated than pre-software mall maps from the 1960s. Moreover, its principal drawing (Fig. 7.12) is conceptually akin to August Merk-Wirz's 1915 "Directory Board" (Fig. 7.1), intended (in the Swiss inventor's words) "to facilitate the finding of certain points on a map, and particularly with respect to that point which corresponds to the location where the selection is made."[88] Contrast this century-old description with the Civix patent's first sentence, whereby "A kiosk is placed on a sidewalk and has stored in an electronic memory the locations of businesses, historical sites, or the like within a predetermined distance of the kiosk"—useful no doubt, but just a mall map.

I would have liked testifying in court, or at least at a formal deposition, but Garmin chose to settle, along with most of the defendants. Financially prudent perhaps, but disappointing. It was nonetheless gratifying to learn later that Microsoft, with deeper pockets and confidence in its position, chose to fight and ultimately prevailed, along with five co-defendants. I was pleased when the District Court not only granted a favorable "summary judgment on grounds of non-infringement" but also awarded the defendants "appropriate costs"—having to pay their own attorney fees would have made this a hollow victory.[89] I was doubly gratified when the federal Court of Appeals upheld the lower court in August 2001.[90]

If the GeoTag and Civix cases prove anything, it is the inability of the patenting system to maintain strict standards of novelty and non-obviousness: a failing akin to editorial laxity in the conventional academic-scientific-

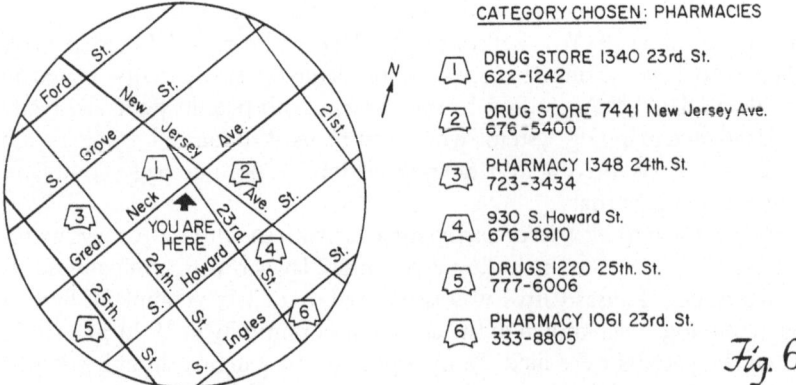

CATEGORY CHOSEN: PHARMACIES

1. DRUG STORE 1340 23rd. St.
   622-1242

2. DRUG STORE 7441 New Jersey Ave.
   676-5400

3. PHARMACY 1348 24th. St.
   723-3434

4. 930 S. Howard St.
   676-8910

5. DRUGS 1220 25th. St.
   777-6006

6. PHARMACY 1061 23rd. St.
   333-8805

Fig. 6

Fig. 7.12 Principal drawing in the first of two patents featured in *Civix v. Microsoft* described what is essentially a mall map (US Patent 4,974,170; 1990)

technical literature, the growth of which seems out of proportion to any real measure of cumulative innovation. Both these parallel literatures have been stressed by an explosion in content in recent decades, and for better or worse, both reward persistence.

NOTES

1. A convenient though not definitive marker is Waldo Tobler's seminal article "Automation and Cartography" in the *Geographical Review* 49 (1959): 526–34. Also see Timothy F. Trainor, "Electronic Map," in *Cartography in the Twentieth Century* (Vol. 6 of *The History of Cartography*), ed. Mark Monmonier, 808–11 (Chicago: University of Chicago Press, 2015).

2. The six patents in my core group, identified by number, inventor's last name, and year of issue, are 1,132,108 (Merk-Wirz 1915), 1,409,894 (Gatliff 1922), 1,929,759 (Swett 1933), 1,932,909 (Pollard and Pollard 1933), 2,876,562 (Stieber 1959), and 3,097,418 (Valliere 1963).

3. Pio Papini, "Indicator for Illustrating and Signaling the Route of a Vehicle," US Patent 1,112,086, filed 13 March 1911, and issued 29 September 1914; and Henry A. Hubschmitt, Jr., "Route Indicator for Automobiles," US Patent 1,236,565, filed 19 October 1916, and issued 14 August 1917. As chapter three noted, several other inventors of automated road guides used electric backlighting, electrically operated bells or buzzers, or electrically synchronized scrolls. These inventors include George Boyden, Max Bremsy, Lee Sherman Chadwick, Howard Cranmer, George Deardroff, Frank J. Lindenthaler, John Protz, and Edward Siegel.

4. The 11 patents, by number and year of issue, are 740,280 (1903); 1,192,829 (1916); 1,339,942 (1920); 1,362,939 (1920); 1,450,335 (1923); 1,512,598 (1924); 1,522,048 (1925); 1,630,916 (1927); 4,998,752 (1991); 5,732,978 (1998); and 7,611,602 (2009).

5. August Merk-Wirz, "Directory-Board," US Patent 1,132,108, filed 18 June 1913, and issued 16 March 1915.

6. Sidney F. Smith (examiner) to B. Singer, 26 November 1913; and Edwin L. Powell, "Indicator for Hotels and Other Places," US Patent 998,391, filed 4 April 1911, and issued 18 July 1911.

7. B. Singer to Commissioner of Patents, 4 May 1914.

8. Cecil A. Keating, "Method of and Apparatus for Recording the Location and Character of Business," US Patent 486,192, filed 16 March 1892, and issued 15 November 1892.

9. "In the Matter of Merk-Wirz's Patent," *Reports of Patent, Design, and Trade Mark Cases* 41.3 (1924): 107–8; according to this account, the patent's title was "Tablet for Use in Ascertaining the Position of Particular Places." Also see August Merk-Wirz, "Maps" [Great Britain Patent 10,189, issued 30 April 1912] in *Patents for Inventions: Abridgements of Specifications, Class 146 (ii), Stationery, Wafers and Seals, Educational Appliances, and Ciphers and Codes, 1909–15*, 88–89 (London: His Majesty's Stationery Office, 1923); and "High Court of Justice, Chancery Division: Claim for Revocation of a Patent, in re August Merk-Wirz Letters Patent," *The Times* [London], 2 May 1923, 5.

10. Robert Gatliff, "Directory," US Patent 1,409,894, filed 21 June 1931, and issued 14 March 1922.

11. See, for example, "Another Gatliff Window Display," *Southern Hardware and Implement Journal* 83.5 (15 April 1920): 24. As for earlier chapters in this book, Ancestry Library Edition was a useful portal to official documents: in Gatliff's case for the Social Security Death Index record of his death in Miami in 1965 as well as for manuscript data from the 1920 and 1930 federal censuses and the 1945 Florida census.

12. Newswire services circulated illustrated stories about Gatliff's designs well beyond Miami. For examples, see "Young Hardware Clerk Becomes Millinery Designer," *Eugene [OR] Guard*, 20 June 1937, 18; and "Hats of Hardware," *Toledo Blade*, 5 March 1951, Pictorial section, 10.

13. A. R. Townshend (examiner) to Franklin H. Hough, 8 July 1921.

14. R. L. Gatliff, "Advertising Device," US Patent 1,764,834, filed 5 July 1929, and issued 17 June 1930.

15. City contract 54625, 14 August 1930, in *Proceedings of the City Commission, City of Grand Rapids, Mich., May 5, 1930 to April 30, 1931, Inclusive*, 208.

16. Edward R. Swett, Jr. and Associates, "Overcoming Obstacles in Fire Alarm Systems" (Muskegon, MI, [1930]). This ten-page booklet was included with Young & Young (attorneys) to Commissioner of Patents, 14 July 1931.

17. Young & Young (attorneys) to Commissioner of Patents, 14 July 1931.

18. Edward R. Swett, Jr., "Fire Alarm System," US Patent 1,929,759, filed 25 August 1930, and issued 10 October 1933.

19. In particular, see M. K. Peck (examiner) to Young & Young, 26 January 1931 and 3 December 1931.

20. M. K. Peck (examiner) to Young & Young, 7 July 1932.

21. This conclusion is based upon the company's publications as well as a search of patents filed by or assigned to Gamewell. For an early corporate history, see Gamewell Fire Alarm Telegraph Company, *Emergency Signaling* (New York, 1916).

22. Oddly, two of the patents in the found set (1,503,439 and 1,991,718) did not include *map* in the text, nor was a map apparent in their drawings. The third (1,251,666), which referred to a traffic-control system, lacked *map* in the text but included a map (but not an electric map) in one of its drawings. The fourth (1,850,177), patented by Oscar Levy, regarded a map as useful but not essential. See note 42.

23. Nathaniel Banks Cregier, "Signal System," US Patent 1,274,514, filed 5 January 1911, and issued 4 January 1918.

24. Nathaniel Banks Cregier, "Signaling System," US Patent 915,075, filed 16 March 1904, and issued 16 March 1909 [five years later].

25. A letter suggesting editorial changes was followed by letters of rejection dated 16 March 1909, 9 February 1911, 13 May 1911, 6 February 1912, 17 March 1913, 20 April 1914, and 7 May 1915. The case file's "Contents" list, which describes the latter as a "final rejection," included additional admonishments dated 6 November 1911 and 3 August 1915.

26. William A. Kirman (examiner) to Nathanial B. Cregier (c/o Hazard & Strause, attorneys), 9 February 1911.

27. Hazard & Strause to Commissioner of Patents, 13 March 1911.

28. C. D. Backus (examiner) to Hazard & Strause, 13 May 1911.

29. William A. Kirman (examiner) to Hazard & Strause, 7 May 1915.

30. Hazard & Strause to Commissioner of Patents, 1 July 1915.

31. Kirman to Hazard & Strause, 3 August 1915.

32. Hazard and Strause to Commissioner of Patents, 21 December 1915.

33. Thomas Ewing (Commissioner of Patents) to Hazard & Strause, 10 January 1916.

34. W. F. Woollard (chief clerk) via Western Union to Cregier, 24 December 1917.
35. John G. Elliott (attorney) to Commissioner of Patents, 2 January 1918.
36. "Cregier Signal Company ... incorporated," *Western Electrician* 42 (18 January 1908): 73; and "Cregier Signal Company" [advertisement], *Chicago Eagle*, 5 November 1910.
37. "Will Test Alarms at Novel Exhibit," *Los Angeles Herald*, 15 October 1909.
38. "Municipal Affairs: Ask Extension for Cregier Co.," *Los Angeles Herald*, 26 February 1910.
39. "Fire Alarm Systems: 'Cregier' Equipment at Los Angeles," *Insurance Engineering* 21.1 (January 1911): 31–38; quotation on 36.
40. Ibid., 34.
41. For a later example, see the floor plan of the Fire Control Center in Syracuse, New York, which included a city map next to a switchboard with seats for two dispatchers. *The Syracuse Fire Alarm System* (Syracuse, NY: Bureau of Municipal Research, 1955), fig. II opp. 7 in appendix.
42. Oscar C. Levy, "Fire Alarm System," US Patent 1,850,177, filed 11 February 1929, and issued 22 March 1932.
43. In a short, privately published family biography, Banks Cregier's son discussed his father's musical talent in detail but was strangely silent about his formal education. Although he reported a collaboration with electronic inventor and radio pioneer Lee De Forest, I have not been able to confirm the scanty details. Ellsworth Banks Cregier, *One Line of the Cregier Family in America* (Chicago, 1959), esp. 30–35. Also see Gloria C. Emma, *The New York Orphan Who Built Chicago: The Story of DeWitt Clinton Cregier, a 19th-Century American Engineering Genius* (St. Charles, IL, 2011), 36.
44. Nathaniel Banks Cregier, "Police Signal System," US Patent 462,808, filed 28 November 1889, and issued 10 November 1891. Nowhere does the patent mention a map.
45. Three of DeWitt Clinton Cregier's patents were for fire hydrants (33,239, 164,149, and 173,768) and a fourth was for a "Pressure Regulator for Water Supply Mains" (257,557). Banks Cregier also gave his father half the rights to an electrical signaling system linking a railroad engineer and the train conductor (476,873).

46. "New Police Signal and Telephone System," *Electrical Engineering and Telephone Magazine* 13.1 (January 1899): 39–41.

47. His son reported adoptions by "such major cities as Chicago, Cleveland, St. Paul, and Los Angeles." Cregier, *One Line of the Cregier Family*, 35.

48. Emma, *The New York Orphan Who Built Chicago*, 36.

49. Barnet Hodes, *Report of the City of Chicago Law Department for 1940* (Chicago, 1941), 82.

50. Robert F. Pollard and Oscar H. Pollard, "Educational Apparatus," US Patent 1,932,909, filed 11 July 1932, and issued 31 October 1933.

51. M. K. Peck (examiner) to C. A. Snow & Company (attorneys), 28 September 1832; and C. A. Snow & Co. to Commissioner of Patents, 2 February 1933.

52. The patents were 2,995,726, 4,449,941, 4,474,557, and 4,609,359.

53. John McGuire and David G. Bories, "Educational Device for Learning Geographical Names and Locations," US Patent 4,449,941, filed 31 January 1983, and issued 22 May 1984.

54. Mark A. Lemley, "Software Patents and the Return of Functional Claiming," *Wisconsin Law Review* 2013.4 (2013): 905–64.

55. Alastair W. Pearson, "Relief Model," in *Cartography in the Twentieth Century* (Vol. 6 of *The History of Cartography*), ed. Mark Monmonier, 1263–67 (Chicago: University of Chicago Press, 2015).

56. Joseph A. Stieber, "Electrical Method and Means for Making Relief Maps," US Patent 2,876,562, filed 10 March 1955, and issued 10 March 1959.

57. A. Berlin (examiner), Notice of Allowance, mailed 11 December 1958.

58. Harvey M. Sapolsky, *Science and the Navy: The History of the Office of Naval Research* (Princeton, NJ: Princeton University Press, 1990), 23–24.

59. Mark Monmonier, "Canada Geographic Information System" in *Cartography in the Twentieth Century* (Vol. 6 of *The History of Cartography*), ed. Mark Monmonier, 189–90 (Chicago: University of Chicago Press, 2015), and Peter L. Pulsifer, "Electronic Cartography: Data Capture and Data Conversion," in *Cartography in the Twentieth Century* (Vol. 6 of *The History of Cartography*),

ed. Mark Monmonier, 366–70 (Chicago: University of Chicago Press, 2015).

60. Sapolsky, *Science and the Navy*, 51, 59.

61. According to the Social Security Death Index, the Joseph A. Stieber who was living in New York State when he applied (before 1951) for his Social Security card was born on 2 July 1910 and was a resident of Fort Myers, Florida, when he died on 10 September 2006. Around 2003 he married Elizabeth Ann Dey Francis, who was born on 2 August 1927 and died on 2 October 2012, in Fort Myers. See "Francis, Elizabeth Ann Dey" [obituary], *Annapolis (MD) Capital,* 25 October 2012, A13. I found no record of a previous spouse for Joseph.

62. Joseph A. Stieber and John B. Weldon, "Map Projections Demonstrator," US Patent 2,932,907, filed 16 January 1956, and issued 19 April 1960; and Joseph A. Stieber and John B. Weldon, "Methods for Forming a Color Impregnation of Transparent Geometrical Shapes," US Patent 3,077,040, filed 16 January 1956, and issued 12 February 1963. The inventors had originally filed a single patent, which was divided early in its examination.

63. Edward G. Valliere, "Electrically Coded Terrain Model Map," US Patent 3,097,418, filed 20 November 1958, and issued 16 July 1963.

64. Edward G. Valliere, "Optical Catheter Means," US Patent 3,267,932, filed 13 November 1962, and issued 23 August 1966; and Edward G. Valliere, "Tunable Air Coil Inductor," US Patent 3,987,386, filed 18 April 1975, and issued 19 October 1976.

65. La Salle University, "Explorer 1965" (1965 Yearbook), Book 22, http://digitalcommons.lasalle.edu/explorer/22; and La Salle University, "The One Hundred and Second Commencement 1965" (1965), La Salle Commencement Programs, Book 34, http://digitalcommons.lasalle.edu/commencement_programs/34.

66. Before it closed in 1996, the Navy Yard employed 7,000 people. Federation of American Scientists, Military Analysis Network, "Philadelphia Naval Shipyard," 7 March 1999, http://fas.org/man/company/shipyard/philadelphia.htm (accessed 21 May 2016).

67. John B. McHugh, "Coordinate Positioner," US Patent 2,981,123, filed 26 November 1958, and issued 25 April 1961; John B. McHugh, "'Z' Axis Drive System," US Patent 2,991,663, filed

26 November 1958, and issued 11 July 1961; and George C. Hand, Jr., and Jacob H. Kubanoff, "Contour Data Recording System," US Patent 3,040,306, filed 4 December 1958, and issued 19 June 1962.

68. W. A. Wiltz (examiner) to J. A. O'Connell and Lawrence S. Epstein (attorneys), 17 July 1959.

69. Epstein to Commissioner of Patents, 21 January 1960; Wiltz to O'Connell & Epstein, 4 August 1960.

70. Wiltz to O'Connell & Epstein, 19 September 1961.

71. Epstein to Commissioner of Patents, 26 October 1962.

72. Wiltz to O'Connell & Epstein, 4 May 1962.

73. Epstein to Board of Appeals, Commissioner of Patents, 26 October 1962 and 11 December 1962. The contested claim, which was claim 5 in the original filing, emerged as claim 1 after renumbering. The changes consisted of making *photoetching* a hyphenated word (in an amendment following the second rejection) and adding the phrase "said contour map having elevation contour areas" (in an amendment following the third rejection).

74. Steven M. Manson, "Computer, Digital," in *Cartography in the Twentieth Century* (Vol. 6 of *The History of Cartography*), ed. Mark Monmonier, 269–70 (Chicago: University of Chicago Press, 2015).

75. The Ngram Viewer is online at https://books.google.com/ngrams. Its operation is described in Jean-Baptiste Michel, Yuan Kui Shen, Aviva Presser Aiden, Adrian Veres, Matthew K. Gray, The Google Books Team, Joseph P. Pickett, Dale Hoiberg, Dan Clancy, Peter Norvig, Jon Orwant, Steven Pinker, Martin A. Nowak, and Erez Lieberman Aiden, "Quantitative Analysis of Culture Using Millions of Digitized Books," *Science* 331 (2011): 176–82.

76. Before 1920 instances of *electric map* in Google Books occasionally referred to maps of an electric railway network or to maps showing areas with residential electric service or electric streetlights.

77. Bradford A. Moore, Alexandre Carlhian, Edouard D. Godfrey, Guillaume Borios, Albert P. Dul, Marcel van Os, and Woo-Ram Lee, "Mapping Application with Interactive Dynamic Scale and Smart Zoom," US Patent 20140365934, filed 12 November 2013, and issued 11 December 2014; and Lanny D. Natucci, Jr.,

Janet L. Schneider, Mark A. Inderhees, Robert R. Dufalo, Jonathan M. Kay, Cristina del Amo Casado, Sanjib Saha, Fernando Gonzalez, and Priyanka B. Vegesna, "Reduced Power Location Determinations for Detecting Geo-fences," US Patent 20140370909, filed 14 June 2013, and issued 18 December 2014.

78. Davide Castelvecchi, "Physics Paper Sets Record with More than 5,000 Authors," *Nature*, 15 May 2015), doi:10.1038/nature.2015.17567, http://www.nature.com/news/physics-paper-sets-record-with-more-than-5-000-authors-1.17567 (25 May 2016).

79. For a concise history of *patent troll*, see Ronald S. Katz, Shawn G. Hansen, and Omair Farooqui, "Patent Trolls: A Selective Etymology," *IP Law 360*, 20 March 2008, http://manatt.com/uploadedFiles/News_and_Events/Articles_By_Us/patentroll.pdf (25 May 2016).

80. For a broader, less biased view of patent trolls, see Colleen V. Chien, "Of Trolls, Davids, Goliaths, and Kings: Narratives and Evidence in the Litigation of High-Tech Patents," *North Carolina Law Review* 87 (2009): 1571–1616. For discussion of possible remedies, see Randall R. Rader, Colleen V. Chien, and David Hricik, "Make Patent Trolls Pay in Court," *New York Times*, 4 June 2013; and "A.G. Schneiderman Announces Groundbreaking Settlement with Abusive 'Patent Troll'" [press release], 14 January 2014, New York State Attorney General's Office, www.ag.ny.gov.

81. Peter D. Dunworth, John W. Veenstra, and Joan Nagelkirk, "Internet Organizer for Accessing Geographically and Topically Based Information," US Patent 5,930,474, filed 31 January 1996, and issued 27 July 1999.

82. Peter Bright, "Google and Microsoft Team Up to Battle Geotagging Patent Troll," *Ars Technica*, 3 March 2011, http://arstechnica.com/information-technology/2011/03/google-and-microsoft-team-up-to-battle-geotagging-patent-troll/ (25 May 2016)

83. Loren Steffy, "Patently Unfair: Barack Obama and Antonin Scalia Agree on One Thing: the Small East Texas Town of Marshall May Be the Worst Thing That Ever Happened to Intellectual Property Law," *Texas Monthly* 42.10 (October 2014): 52–60.

84. Bright, *op. cit.*

85. *Microsoft Corp. v. Geotag, Inc.*, No. 2015-1140 (Fed. Cir. Apr. 1, 2016). The decision lists Google Inc. as a plaintiff-appellee, and includes a concise summary of the patent as well as a procedural history of the case. For the district court decision, see *Microsoft Corporation and Google Inc. v. Geotag Inc.*, 847 F.Supp.2d 675 (2012). There's more: several months before the Circuit Court announced its decision Google and Microsoft agreed to drop pending lawsuits against each other. A new ethos might stifle patent litigation, much of which is not only pointless but a hindrance to technological innovation. Shira Ovide and Alistair Barr, "Microsoft, Google Agree to Dismiss All Pending Patent Infringement Lawsuits," *Wall Street Journal*, 30 September 2015.

86. *CIVIX-DDL, LLC v. Microsoft Corporation et al.*, case 99-B-172, filed 26 January 1999, in the Federal District Court for Colorado. For a fuller discussion of my involvement, see my *Adventures in Academic Cartography: A Memoir* (Syracuse, NY: Bar Scale Press, 2014), 148–49, 167–68.

87. W. Lincoln Bouve and Edward Holmes, "Electronic Directory for Identifying a Selected Group of Subscribers," US Patent 4,974,170, filed 25 January 1990, and issued 27 November 1990; and W. Lincoln Bouve, William T. Semple, and Steven W. Oxman, "System and Method for Remotely Accessing a Selected Group of Items of Interest from a Database," US Patent 5,682,525, filed 11 January 1995, and issued 28 September 1997.

88. August Merk-Wirz, "Directory-Board," US Patent 1,132,108, filed 18 June 1913, and issued 16 March 1915.

89. *Civix-DDL v. Microsoft Corp., et al.*, 84 F. Supp. 2d 1132; 2000 U.S. Dist. LEXIS 717 (24 January 2000).

90. *Civix-DDL v. Microsoft Corp., et al.*, 18 Fed. Appx. 892; 2001 U.S. App. LEXIS 19597 (22 August 2001).

Appendix: How to Find a Patent

According to the *University of Chicago Manual of Style*, the proper bibliographic citation for a patent includes the inventor(s), the patent's title and document number, and the dates on which the patent was filed and issued. When a patent is issued (always on a Tuesday), it is assigned a unique document number, which is the most reliable way to find (and download) a patent document, find additional information that is not in the patent document, or retrieve the case file at the National Archives. By contrast, patent titles are not unique and are often uninformative; for example, the title of John Byron Plato's patent for a clever rural address system is merely "Map or Chart." Be wary that for older patents, the inventor's name noted on the first page of the patent might be less complete than the inventor's name printed at the beginning of the specification. In Plato's case, he is identified as "J. B. Plato" at the top of the first drawing page, as "John Bryon Plato" at the top of the first text page, and as "John B. Plato" at the beginning of his personal statement, copied directly from the application. The patent document also identifies the inventor's place of residence, which can be helpful when using Big Microdata tools like Ancestry.com to explore the inventor's occupation, education, training, and dates of birth and death, which are especially useful if two people different in age share the same first and last names.

If the patent number is known, the complete patent document is readily downloaded from the Google Patents database. Just type the patent

© Mark Monmonier 2017
Mark Monmonier, *Patents and Cartographic Inventions*,
Palgrave Studies in the History of Science and Technology,
DOI 10.1007/978-3-319-51040-8

number into the search box at patents.google.com, wait for the search result, and request a PDF download. The search result lists earlier patents cited by the examiner and provides links to "similar documents"—journal articles as well as patents—for which the selection criteria are not readily apparent. The US Patent and Trademark Office (USPTO) also delivers complete patent documents at its online Patent Full-Text and Image Database (http://patft.uspto.gov/netahtml/PTO/srchnum.htm), but the process is less straightforward.

Google Patents also provides an advanced search (https://www.google.com/advanced_patent_search) based on keywords, patent number, title, inventor, original assignee, classification (current US, international, or "cooperative"), and date (filing date or issue date). Results can be restricted to a time period specified by start date, end date, or both. Keyword searching, which the USPTO also supports, is thwarted by uninformative titles and a proliferation of synonyms, which add tedium and uncertainty.

Unless you seek patents issued to a particular inventor, classification search is usually more reliable than keyword search, provided you explore the classification sufficiently thoroughly to find all relevant categories. An added complication is the dynamics of patent classifications, which must reflect changing technology and are used mostly by patent examiners. As noted in Chap. 1, the US Patent Classification (USPC), which I used for this book, has been revised on several occasions. Updating stopped in 2015, when the USPTO officially adopted the Cooperative Patent Classification (CPC), a joint venture of the USPTO and the European Patent Office. The CPC incorporates the more reliable aspects of the USPC and the European Classification System, a refinement of the International Patent Classification (IPC) system, initiated in 1971. Although the USPC was mothballed, it is still available at the USPTO website and remains eminently useful for historical research.

Because most inventions are assigned to multiple categories, a suitably robust set of representative inventions can be a useful starting point for trial-and-error exploration of a classification's structure and definitions. Plato's rural directories are a useful example insofar as they included the patent number (1147749), which I typed into the query box at the aforementioned USPTO Patent Full-Text and Image Database, which revealed the patent's sole "Current U.S. Class" as 283/34. I then pulled up the USPTO's Patent Classification homepage (http://www.uspto.gov/web/patents/classification/), selected USPC as the classification system, entered 283 and 34 as the two-part "classification symbol," chose "Definitions" as the desired content, and clicked Submit. The result identified the category as Printed matter/Maps, defined the subclass Maps by saying "The indicia delineate geographical features," and defined the class Printed matter with a suitably border and more cumbersome explanation that does not warrant repeating.

Just below Printed matter/Maps, the search result shows the indented category Printed matter/Indexed maps, coded 283/35, with the subcategory defined by "the indicia involving means which facilitate finding some of the geographical features." Oddly, this is not a cross-reference category for Plato's rural address.

To retrieve a list of all patents assigned to a particular category, I recommend the USPTO Patent Full-Text and Image Database (PatFT), online at http://patft.uspto.gov/. Full-text information is provided for patents issued since 1975, but for earlier patents only the patent's number, classification, and issue date are available. To find all patents in either the 283/34 or 283/35, I clicked on Advanced Search; typed "CCL/283/34 or CCL/283/35" into the Query box; set Select Years to "1790 to present"; and clicked Submit. (CCL, the "field code" for the "Current US Classification," still refers to the USPC. The other Select Years option is "1976 to present," which searches only patents for which full-text documents are available.) The result was the first 50 of 304 patents, sorted from newest to oldest and identified by number and title.

Although each item in the result is a hyperlink to information about the patent, I prefer to use the insertion bar and Shift key to copy-paste all 50 lines into an Excel workbook, use the same technique to capture information for the other 254 patents (in 6 groups: 51–100, 101–150, 151–200, 201–250, 251–300, and 301–304), remove all the hyperlinks, and copy-paste patent numbers individually into Google Patents. Although the USPTO lists do not provide titles for the older patents, the hyperlinks lead to classification for all three systems (USPC, CPC, and IPC). As noted in Chap. 1, patents are often assigned to more than one category. Indeed, 26 of the 304 patents in the found set were assigned to both 283/34 and 283/35.

INDEX